the DNA question

Where Does The Information Come From?

by Stephen Orla Searfoss

Copyright © 2019 Stephen O. Searfoss.

All rights reserved. No part of this publication may be reproduced, distributed, or transmitted in any form or by any means, including photocopying, recording, or other electronic or mechanical methods, without the prior written permission of the publisher, except in the case of brief quotations embodied in critical reviews and certain other noncommercial uses permitted by copyright law. For permission requests, write to the publisher, addressed "Attention: Permissions Coordinator," at the address below.

ISBN: 9781697012446 (Paperback)

Library of Congress Control Number:2019919767

ASIN: B07YN1KPD1 (Kindle eBook)

Ordering Information:
This book can be ordered through Amazon.com.

Imprint: Independently published in the United States of America.

First Edition 2019.

Stephen O. Searfoss
1 Pinon Cove
Cedar Crest, New Mexico, 87008
Stephen@theDNAquestion.com

www.theDNAquestion.com

Manuscript v1.06 January 1, 2021

Dedication and Acknowledgements

I dedicate this book with love to Becky, my wife and my lifelong companion.

My good friend of many decades, Stan Harder, told me I needed to write a book to get my ideas into a clear argument. I value his advice enough to write The DNA Question, which has turned out to be about a year-long project. Thanks Stan. I want to thank Michelle Scholz and Jaclyn R. Holder for partnering with me to get this DNA Information Research project started. I thank you both for your support. Thanks, Michelle, for giving me so many hours and days of invaluable editing advice. Jaclyn, thank you for your suggestions encouragement.

Three people invested many hours of their time reading the book and offering criticisms and edits: Professor Abraham Martinez Baini, from the Autonomous University of Queretaro in Mexico, Daniel Calzada, a co-worker in information technology, and Samuel Salcedo Gonzalez data science consultant. Thank you, Abraham, Daniel and Sam. Dr. Jorge Yañez, an oncologist, gave criticisms and suggestions. Thank you, doctor Jorge. I also want to thank Adam Hernandez, Joy Ost, Alejandro Alonso, and others who prefer to remain anonymous, read the manuscript and given me feedback, suggestions, and

encouragement,

My son Steve took the almost final manuscript and worked with me to highlight what the book is all about and get it in dynamic form. Thank you, son.

I have to thank the many scientists who have made their careers and lives a search for knowledge about life, genetics, DNA, and information. To name them would take more space than this whole book. A sampling of these would be Gregor Mendel, who was willing to keep pea plants from random cross-pollination and meticulously count thousands of pea plants. David Morgan and his crew, who radiated tens of thousands of flies and then examined each of them one by one looking for mutations. Lastly, a young Greek electronics professor who began to teach me about the atomic world where electrons are repelled from one atom and attracted to another atom. I am thankful that I got to have a childhood where I could play around with dangerous chemicals, electricity, plants, and wild animals.

Stephen O. Searfoss

Author's contact information:
Stephen@theDNAquestion.com

Contents

Preface	8
Introduction	11
PART 1 - The Marvelous Functionality of Life	**14**
Chapter 1 - Marvelous Functionality and Complexity of Organisms	15
Chapter 2 - Where is the "know-how" in cells?	19
Chapter 3 - Information in DNA	24
Chapter 4 - Abstract Information	
PART 2 - The Source of Information	**42**
Chapter 5 - Intelligent Agents Add and Change Information in DNA	49
Chapter 6 - The Evidence is in the Lab	57
Chapter 7 - What Else Changes DNA Information?	62
Chapter 8 - Relationship Between Mutations and Information in DNA	73
Chapter 9 - Does Environment Change DNA?	79
Chapter 10 - Does The Environment Send Information To DNA?	83
Chapter 11 - Does Natural Selection Change DNA?	101
Chapter 12 - What Is Natural Selection Supposed To Do?	109
PART 3 - Some Other Explanation?	**120**
Chapter 13 - Do Epigenetics Change DNA?	121
Conclusion	130
Footnotes	134
Glossary	144
Image and Illustration Credits	152
Additional Resources	156
Appendix - Abstract Information	157

Preface

I have always been interested in how information works and what it can and cannot do. I have spent 50 years as a professional in the field of information technology (IT). Most of my career has been designing computer programs, but I have also designed electronic and mechanical interfaces from computers to machines. My work has been to use information in the real world: take specific inputs, add instructions, and produce specified outcomes.

Some ten years ago, I became fascinated with biology, genetics, and DNA – which contains another type of information. I started to study how DNA works and saw that something did not add up. From what I know about information and systems, I did not see how DNA could come to contain the information it has. I went searching for explanations, and every explanation I analyzed came up short, which is why I decided to write this book.

What I offer you is a look at DNA from the perspective of an information technologist.

There are two things from my experience in IT that are pertinent to the subject of information in DNA. As a software architect, I listened to what people wanted to do with a computer and translated those needs into pieces of data, processes, and functionality. Consequently, whenever I read about biological processes and organisms, I automatically begin to see which data and which process controls are needed to produce these processes, and functionality.

The other experience involves explaining information technology to those who have little or no knowledge of IT. Whenever I was asked to design a software solution for someone, in the interest of being honest, I would first meet with the client and go over what they wanted the software to do. Then I would write a first draft of the proposal. In these proposals, I would list the data we were going to handle. For each piece of data, I would define what type of data it was and specify its characteristics. I would also explain the functions or processes that would be performed with this data. This proposal also detailed what they would be able to see on different screens and reports. Usually, the client would carefully study this draft, and then we would meet again to make changes and additions to these specifications. After a few iterations, they would accept that the document described what they wanted the new software to do.

Once they thought the proposal was complete and accurate, I would ask them to sign my disclaimer. This disclaimer explained that the software I was going to develop would only store and manipulate the data specified in the proposal. The disclaimer clarified that the only processes or manipulation of data would be what the document described. Likewise, they would see only the screens and reports described in the document. Almost everyone would balk at signing this disclosure. There was one company that could never settle on the limits of what they wanted the software to do. Generally, they harbored the idea that the computer and its software were almost magical. They imagined that the software would do many other things that were not to be input into the information system of the software.

What does this have to do with the information in DNA? Pretty much the same thing. I am going to keep us honest. All functionality in organisms has to be in the data and processes around the data. There is no magic; if there is no information and processes to produce functionality and form in organisms, then they will not have that functionality and form. Or if we look at this from the other side, if there are functionality and form, somewhere, in some format, in some medium, there has to be the information needed to specify that function and form. There is no magic.

When we want to learn about living organisms, we

automatically assume that the proper means of approaching this would be biology. After all, the definition of biology is *"a branch of knowledge that deals with living organisms and vital processes."* [27] It has been shown, over the last 70 years, that DNA contains information that describes and determines the characteristics and functions of organisms. In this book, we will take a step away from most of the biological details, such as the chemical characteristics of organisms, and instead consider DNA as a repository of information. From the perspective of this repository, what is the source of its information?

Is it valid to look at DNA from an information perspective? Which is more critical, the biochemical details of the processes, or the information? I will argue in this book that the information is most important as it determines the makeup of the fundamental parts of organisms.

> *DNA sequences provide the "basis for the genetic code for all of the proteins synthesized by our bodies, and these, in turn, provide the basis for the structure of all of our cells, all of our enzymes, and all of our inherited traits and characteristics"* [23]

I intend to show the shortcomings of accepted theories, and ask: where is the source of the information that is in this DNA repository?

The question about the information in DNA is not just an interesting question, it is *the* question because if we can answer it, we can know the origin of life.

Jaques Monod, a Nobel Prize recipient, considered by many as one of the founders of molecular biology, explained the importance of the DNA question:

> *"Consequently no other science has quite the same significance for man; none has already so heavily contributed to the shaping of modern thought, profoundly and definitively affected as it has been in every domain - philosophical, religious, political - by the advent of the theory of evolution."* 31

I agree with Monod. Our understanding of the origin of life affects every domain of humanity. If our explanation for the origin of life is not valid, then our philosophy, religion, politics, and every domain will have a bad foundation. Scientific investigation has narrowed *the* question to our question: where does the information in DNA come from?

Introduction

When most people think about the origin of life and the process by which it comes into being, they think about evolution and DNA.

The problem is that evolution fails to explain the origin of the information in DNA.

DNA contains all the information needed to make a complete organism starting with just one cell. Most people take it on faith that evolution added to DNA all of the information needed for this process to unfold. As I have studied this issue, I have come to understand that there is not a solid scientific explanation for this. We believe scientists when they tell us that they understand how DNA works. But the more I have analyzed this, the more I am convinced that their understanding is incomplete. DNA at the end of the day is information. And the accepted understanding of the origin of information in DNA is not based on sound science.

DNA contains detailed information on how to make all the proteins of an organism. The current scientific explanation for the origin of this information is that natural selection filters this information out of a flow of random mutations. The problem with this explanation is that, because natural selection filters based on reproducibility, it filters based on function. But you need to have information to generate functionality. Yet, there is currently no explanation proposed of how this necessary information was added to DNA.

We also do not understand how DNA contains the information that determines how to build parts larger than proteins, such as cells, organs, organ systems, body parts, and shapes. Our current theory of genetics says that DNA contains all this information. Yet, there is currently no explanation of where and how this information is encoded in DNA. Proposing something with no explanation is bad science. To think that these theories of origins have explained who we are as humans is unscientific. What this means is that the question about how life comes into being is still very much an open question.

This book is divided into three parts. Part 1 is an overview of the marvelous functionality of life and the incredibly complex information contained in DNA that is at the heart of this marvelous functionality. Readers who are already familiar with how DNA works may want to skip this section and begin by reading Part 2.

DNA is information, and it is incredibly complex, and we don't know where this complex information comes from. Scientists have offered three possible explanations for where this complex information comes from: mutations, the environment, and natural selection. In Part 2, I look at each of these three possible explanations in detail and explain why none of them are satisfactory.

Having no convincing explanation of where this complex information comes from, scientists are reaching for other explanations external to DNA. One of these explanations is epigenetics, which can determine which parts of DNA are active or inactive. I will examine epigenetics in Part 3. I will show how this too, is not a plausible explanation for how DNA came to contain the complex information we see. So we are left with the question: what is the source of the information behind the marvelous complexity of life?

Get updated information on our page
www.theDNAquestion.com

www.theDNAquestion.com

PART 1

The Marvelous Functionality of Life

Chapter 1
Marvelous Functionality and Complexity of Organisms

Since we are human, let's begin by looking at how a human organism starts its life. Every human being starts out as just a single cell, made from an egg and a sperm. What a cell!

This cell divides into two cells, and the two cells split into four cells. The four cells divide into eight cells and so forth. These first cells are of a type known as *stem cells,* or to be more precise, embryonic stem cells. The only function of embryonic stem cells is to reproduce. They are able to produce any type of cell. At different times these stem cells begin to produce diverse types of cells. In a human body, there are around 200 different types of cells, yet the human body starts with just one embryonic stem cell. Each type of cell is different and has a unique function. Some of the cells that split off from this first cell eventually produce cells like T cells, for example. T cells are what we commonly refer to as white blood cells, which play a critical role in our immune system.

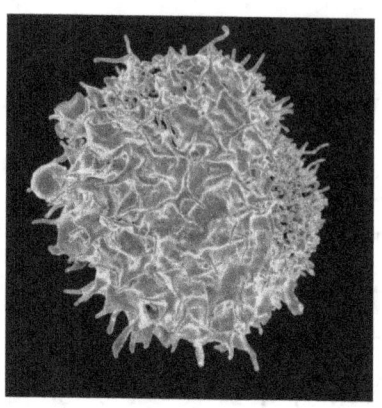

T Cell [1]

Other types of cells produced are *Paneth cells* and the rod and cone cells that you'd find in your eye. How do this first cell and its offspring "know" how to make all these different types of cells? Two hundred types of cells, yet all the "know-how" was there in that first cell. This process continues until it has produced 30 to 40 trillion, cells of these 200 or so types. How do these cells "know" when, where, and how to produce these so that they make a human being?

This first cell does not just "know" how to produce trillions of diverse types of cells, as the organism, in this case, a human grows and develops, these cells form themselves into *organs*. Organs such as a heart, a stomach, or lungs. How do these cells "know" how to develop into organs?

These different organs interconnect and work

together to build about ten different organ systems. We call these organ systems as they consist of two or more organs functioning together. An example is the circulatory system. The circulatory system includes, among other things, the heart, blood, and blood vessels. The digestive system includes the mouth, the esophagus, the stomach, the intestines and others.

How did that one cell "know" how to produce all these different types of cells? All of these various organs? How does it "know" how to make these interdependent organ systems?

Think about the skeletal system, another of the ten organ systems. There are some 206 different bones in the human body. Each one of these 206 bones has a unique shape and characteristics.

Of course, bones are much more than just their shape. In many bones, there are openings so that blood vessels and nerves can run through them. The surface of a bone is made up of compact bone. Inside there are some spaces filled with spongy bone, others with bone marrow. How do this first cell and its descendants "know" what shape the bone should have? How do these cells "know," which is the space to be occupied by the bone marrow and which areas to fill with spongy bone? Or even how, when, and where to produce the cells needed for spongy bone, marrow, or compact bone. How does this original cell "know" how to do all of this for more than 200 types of bones?

How did this original first cell "know" how to interconnect all these bones in a skeletal system with tendons and muscles to hold it all together?

We would need to read volumes of medical books to appreciate the details of the structure and functionality of all the organs, the organ systems, and the interplay between these. How did this original cell "know" how to multiply and diversify into all these different interdependent organ systems that make up the human body?

What is the origin of the marvelous complexity of life?

Chapter 2
Where is the "know-how" in cells?

Where is the "know-how" that produces this marvelous complexity?

We owe much of our understanding of how this first cell "knows" how to grow into an organism with all its characteristics and functions to Gregor Mendel who applied the scientific method proposed by Roger Bacon, to successive generations of pea plants. He studied how the traits of pea plants are passed down to their offspring. Based on this research, Mendel gave us the basic laws of inheritance. It took a few decades for Mendel's work to be appreciated but now Gregor Mendel is known as the father of genetics.

In the last century and a half, a growing number of dedicated scientists have built on Mendel's observations to bring us to our current understanding of *genetics*. In 1953, *DNA* was discovered to be the container of the "know-how" or information of genetics. Currently it's accepted that DNA contains all the information that describes and determines the characteristics and functionality of all organisms.

We still do not know a whole lot about how this information is stored in DNA. The part of DNA that is well understood is called "coding DNA." Coding DNA contains information necessary to make all the *proteins* of organisms.

There are two usages of the word "protein," so let's clarify. We are not talking protein as a nutrient, as in protein bars or protein-rich foods. No, within biology, a protein is a class of molecules that are the fundamental building block of living things. It takes many different proteins to make a cell, for instance. Proteins are quite varied. There are some 20,000 different proteins that make up our bodies. Proteins are used for transport and storage. They can be *antibodies* and messengers. Proteins are also structural components used, for example, to make wood and bones. We will go into detail on how the information in DNA is used to build proteins, and explain the role of coding DNA.

Let's talk about information in DNA which we still don't know a lot about, before going into detail about coding DNA. What information guides proteins to become cells? And as we asked earlier, what guides cells to become organs and all these organs and cells to become a harmonious organism?

This information has yet to be found and deciphered in DNA, yet it is assumed to be in DNA, as we can see from this quote from a journal article by biologist Terrence Lappin and others.

> *"It is a fascinating thought that the single cell zygote contains all the information required for the development of the adult organism. Understanding how this information is encoded and deciphered is a major uncompleted scientific challenge."* 24

Another cutting edge scientist, Michael Levin, describes what this information needs to be in these terms.

> *"The capacity to generate a complex organism from the single cell of a fertilized egg is one of the most amazing qualities of multicellular creatures."* 25

Michael Levin, explains that we do not know how and where the information needed to guide the building of a basic body plan and defining the structures that will be formed is encoded in DNA. He thinks that successful research will find the information required to direct this *4-dimensional* dynamical system that reliably builds and maintains the complex morphology of organisms.

Did you notice how one of these scientists used the word "information" to describe what we were calling "know-how" in organisms? The "know-how" is derived from information.

A simple way to think about this is that every cell in an organism needs to be of a particular type, in an approximate location at a given moment in time, for the organism to live and function. It is this information that will take proteins and make a turtle with a shell, or with almost the same proteins, a bird with feathers. We are yet to find this information in DNA. Nevertheless, it is thought that DNA contains all this information. At this point, we could say that this is a widely accepted presupposition. After all, if this information is not in DNA, then where is it kept?

DNA may very well contain more information than any single repository produced by humans. I like to make a comparison with a Boeing 747 aircraft. Suppose we had in one reference book or memory stick all the information needed to build a Boeing 747. How much information would we need? But wait, there is an important caveat here. We are not talking about a repository that would specify that we get some carbon fiber components from this supplier, some high-pressure hydraulic hoses from another supplier, and digital computer units from yet another supplier. That would not be a good comparison at all. No, we would need an information repository that contained the entire structure of a 747 with all its functionality, starting only from molecules! Now imagine that repository of information. That is what DNA contains for each organism. Of course, to build a 747 or an organism, you need tools. DNA includes the information needed to make the tools that will be required. You see, DNA contains all the information required to build a human or any other organism from the molecules on up.

Chapter 3
Information in DNA

We have seen how incredibly complex a living organism can be and that all that complexity originates from a single cell. In the case of humans, a single original stem cell "knows" how to build the 200 different kinds of cells that are used to create bones, muscle, organs, eyes and everything else in the human body.

How is this sophisticated "know-how" stored? And how is it transmitted as that first cell multiplies into 30 to 40,000,000,000,000 cells?

To answer these questions, we need a basic simple understanding of coding DNA.

You have probably seen a drawing of DNA, like the one below, that looks like a twisted ladder.

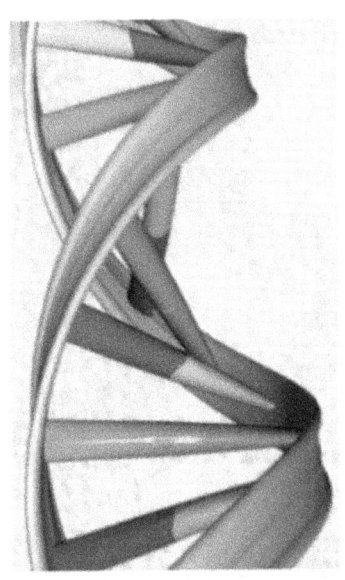

DNA Like a Twisted Ladder [2]

Each one of the steps on this DNA ladder is made of two molecules known as *nucleobases* or just *bases*. There are only four types of bases in the DNA of all organisms. These bases are small molecules consisting of 13 to 16 atoms each. These are referred to simply by the first letter of their names, which are A, C, G, and T. The letter A refers to *adenine,* C refers to *cytosine,* G refers to *guanine,* and T refers to *thymine.* We are looking at this from an information perspective. We don't need to know the names of the bases. The letters A, C, G, and T, are sufficient to know we have four different basic variants of data.

Here is an illustration with details of these bases in

our DNA ladder.

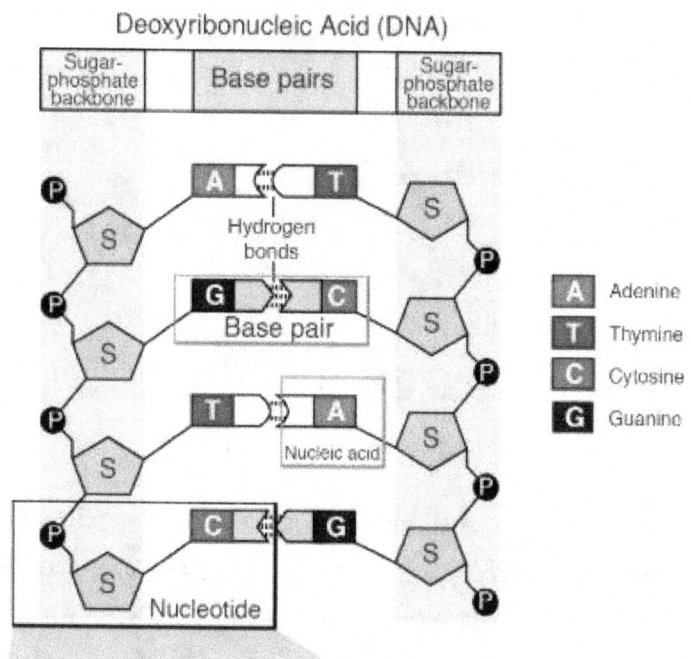

DNA Base Pairs [3]

It is important to note that if A, adenine, is on one side of one of a step of this ladder, then it will always be joined with T, thymine on the other side. Likewise, if there is C, cytosine on one side of a rung on this ladder, then there will always be G, guanine on the other side.

Reading the letters of the types of bases on the

right side of this illustration from top to bottom, you can see how information is stored in these bases. In this case, the information sequence is T, C, A, then G. Note that the information on the left side is different from that on the right side. Since each base on the right side will only join with one type of base on the left side, the left side is a mirror copy of the right side. Copies of DNA are made by unzipping the two sides and making a mirror copy of each side. To get the information, we only need to read the T, C, A, and G off one side of the DNA strand.

Notice in this illustration, a base, with its fragment of the backbone (sugar-phosphate backbone), is called a *nucleotide*. We will refer to these pieces of data as nucleotides.

Before we go any further we need to define a couple of terms that we will use throughout this book. First of all information. When we mention *information* in this book, what we mean is what Merriam-Webster gives as one definition of *information* here:

> *2 b:the attribute inherent in and communicated by one of two or more alternative sequences or arrangements of something (such as nucleotides in DNA or binary digits in a computer program) that produce specific effects* 29

In this chapter we will see how sequences of nucleotides in DNA produce proteins. This is one example of information in DNA, as defined by Merriam-Webster.

We have already introduced the four different bases or nucleotides found in DNA. Each nucleotide is a piece of data, as Merriam-Webster, defines *data* as:

> **2: information in digital form that can be transmitted or processed**
> **3: information output by a sensing device or organ that includes both useful and irrelevant or redundant information and must be processed to be meaningful** 28

Nucleotides are data. Note that the definition mentions that some of the data may be irrelevant. We will see that some of the nucleotides in DNA either do not contain information or we do not know how any information is encoded by them.

The definition of data mentions digital form. We refer to nucleotides with characters, but they are considered digital. Richard Dawkins explains this in his article, The Information Challenge.

> *"DNA carries information in a very computer-like way, and we can measure the genome's capacity in bits too, if we wish. DNA doesn't use a binary code, but a quaternary one. Whereas the unit of*

information in the computer is a 1 or a 0, the unit in DNA can be T, A, C or G." 17

We use the decimal system to express numbers and calculations. The decimal system has ten digits, 0 through 9. Computers use the binary system of numbers which only has two numbers called bits. The binary system only uses the bits, 0 and 1. DNA instead of bits uses nucleotides. Instead of a 0 or a 1, we have A, C, G or T in DNA data.

With these two definitions we can continue looking at information in DNA.

So how is it that coding DNA contains the information on how to make all the proteins of organisms?

Proteins are made up of smaller components, which are amino acids. These amino acids are also connected sequentially one after the other in what is called a *polypeptide* chain. Some proteins consist of a single polypeptide chain. Other proteins are made from several polypeptide chains.

Decades of scientific study demonstrated that the variance of nucleotides in their sequence in DNA determines which amino acids are joined and in what order in polypeptide chains and thus in proteins. So the "know-how" to build proteins is in the information in

DNA!

There is a challenge here. We have explained that there are only four different nucleotides, but there are 20 commonly used amino acids.

NOTE: There are two other amino acids used less frequently.

> *"Selenocysteine (Sec) and pyrrolysine (Pyl) are rare amino acids that are cotranslationally inserted into proteins and known as the 21st and 22nd amino acids in the genetic code. Sec and Pyl are encoded by UGA and UAG codons, respectively, which normally serve as stop signals."* [55]

How can four types of nucleotides specify one of 20 amino acids?

Coding DNA, or to be more precise, protein-coding DNA, has been deciphered. Now, we know that groups of three nucleotides are used to define a specific amino acid. This group of three nucleotides is called a *codon*. Since there are four possible types of nucleotides, three nucleotides give us 64 possible combinations.

The math for this is simple: $4 \times 4 \times 4 = 64$

3 nucleotides = 1 codon ⟶ 1 amino acid

The following table shows which amino acid is specified by each codon or combination of three nucleotides.

1st base		2nd base				3rd base
	T	C	A	G		
T	TTT (Phe/F) Phenylalanine	TCT (Ser/S) Serine	TAT (Tyr/Y) Tyrosine	TGT (Cys/C) Cysteine	T
	TTC	TCC	TAC	TGC	C
	TTA (Leu/L) Leucine	TCA	TAA Stop (Ochre)[B]	TGA Stop (Opal)[B]	A
	TTG[A]	TCG	TAG Stop (Amber)[B]	TGG (Trp/W) Tryptophan	G
C	CTT (Leu/L) Leucine	CCT (Pro/P) Proline	CAT (His/H) Histidine	CGT (Arg/R) Arginine	T
	CTC	CCC	CAC	CGC	C
	CTA	CCA	CAA (Gln/Q) Glutamine	CGA	A
	CTG[A]	CCG	CAG	CGG	G
A	ATT (Ile/I) Isoleucine	ACT (Thr/T) Threonine	AAT (Asn/N) Asparagine	AGT (Ser/S) Serine	T
	ATC	ACC	AAC	AGC	C
	ATA	ACA	AAA (Lys/K) Lysine	AGA (Arg/R) Arginine	A
	ATG[A] (Met/M) Methionine	ACG	AAG	AGG	G
G	GTT (Val/V) Valine	GCT (Ala/A) Alanine	GAT (Asp/D) Aspartic acid	GGT (Gly/G) Glycine	T
	GTC	GCC	GAC	GGC	C
	GTA	GCA	GAA (Glu/E) Glutamic acid	GGA	A
	GTG	GCG	GAG	GGG	G

DNA Codon Table [4]

On the table in the left column, you can see nucleotides A, C, G, and T, as row headers. There you would choose the correct row for the first nucleotide. On the second row from the top, you would select the column that corresponds to the second nucleotide. Finally, on the right side, you pick the row that corresponds to the third nucleotide in the codon. So if we know which nucleotides are in a codon and in what order they appear, we can lookup which amino acid would be added to the polypeptide chain.

Looking at this table, we can appreciate why scientists speak of the information in DNA. It is a data table. This table has information only because there is a biological information system that translates the codon's data to amino acids.

So we have three nucleotides that make a codon, and each codon specifies an amino acid to be added to the polypeptide chain. The codon ATG is the signal to start the building process. It also specifies the first amino acid, Methionine.

Then each amino acid specified will be linked to the previous amino acid to make a chain of amino acids called a polypeptide.

As we saw, there are 64 possible combinations of nucleotides in a codon, so more than one codon can specify the same amino acid. In the DNA Codon table, you can see that there are four codons or nucleotide combinations used for what we can call punctuation. One signals to start building the polypeptide chain and three different codons to stop making the chain. So that leaves 60 additional combinations of nucleotides that could specify 60 unique codons. Each one of these codons specifies an amino acid. There are only 20 common amino acids. So you will notice on this table, there are quite a few amino acids that can be specified by more than one combination of nucleotides.

Coding DNA information is not just a random list of amino acids. The sequence or order of these codons within the DNA chain determines the order in which the amino acids get added to the polypeptide chain. The order of the amino acids in the polypeptide chain is of utmost importance. These amino acids are not just blended as we would mix liquids, for instance. These polypeptide chains will be folded into unique shapes. The sequence of amino acids is critical in determining the shape and function of the protein. The information in DNA produces functioning proteins.

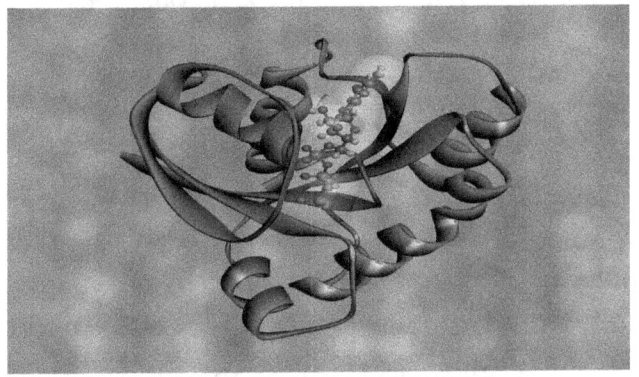

Kras Protein Shape [5]

Three combinations of nucleotides in codons are used to stop building the polypeptide. In exceptional cases, two of the codons which signal to stop building the polypeptide chain are used instead to specify amino acids 21 and 22.

I have explained the relationship between the information in DNA and building of proteins. I must point out that there is an intermediate step in which the information in DNA is transcribed into another chemical form, before being used to build a protein.

DNA is nicely tucked away in the *nucleus* of cells. In the case of humans within the nucleus of every cell, there are two copies of all 22 *chromosomes* and two sex chromosomes of DNA. Proteins are not assembled within the nucleus of the cell.

> *In order to build a protein, an enzyme called RNA polymerases is sent into the nucleus of the cell to transcribe the DNA sequence of the desired protein into what is called messenger RNA or mRNA.* 40

mRNA is different than DNA in several ways. First of all, it does not have the same twisted ladder structure as DNA. It looks more like one side of this ladder, and it is not as straight as a side of DNA. Instead of using T, thymine, RNA uses U, *uracil.* When DNA information is passed over to RNA, the Ts are transcribed as Us. Below is a diagram of DNA being transcribed into RNA and then transported out of the nucleus.

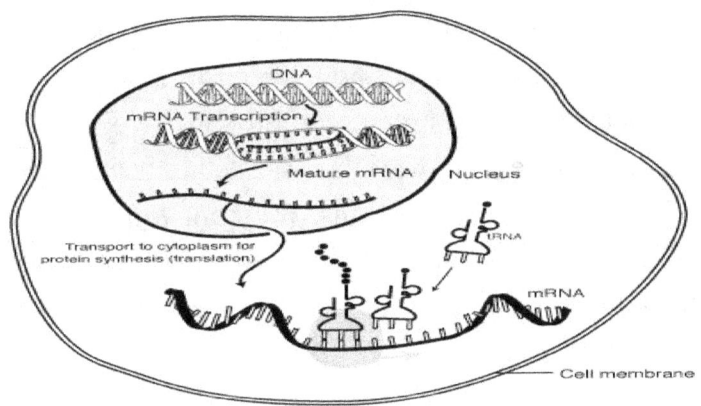

DNA to mRNA [6]

In this translation process from DNA to RNA, we can appreciate that what is critical is the information and not the medium used to represent it. The information passes from its chemical representation in DNA and is represented now in a different chemical representation in RNA. The information and what it represents remains the same.

> *"The first step a cell takes in reading out a needed part of its genetic instructions is to copy a particular portion of its DNA nucleotide sequence — a gene — into an RNA nucleotide sequence. The information in RNA, although copied into another chemical form, is still written in essentially the same language as it is in DNA the language of a nucleotide sequence. Hence the name transcription."* 1

With these explanations, we give you a simplified yet accurate description of how the information in DNA determines the structure of proteins. We're taking a bird's eye view of this. We're not trying to teach you all the details of this. We want you to have enough knowledge and technical definitions so that we can evaluate the most common suggested sources for this information.

From an information perspective, there is nothing magical occurring here. There is no shortcut, no way to encode 20 different types of amino acids with less than three nucleotides. Nor is there any way to have the amino acids in the needed sequence without having them specified in that sequence in the DNA. The basis of materialistic science is that every effect must have a cause. Every amino acid needs to be specified by three nucleotides.

What is meant by every effect has to have a cause? Let's review. Cells are built from proteins. Proteins are made of polypeptide chains. Polypeptide chains are made of specific sequences of amino acids. Specific sequences of amino acids must be stipulated by a specific sequence of codons or nucleotides.

4 nucleotides ⟹ 20 amino acids ⟹ 20,000 proteins ⟹ 200+ cell types
⟹ **Information flows one way** ⟹

The transmission of DNA information during this process must be very exact, as we can see in this example taken from the online notes of a biology course. It gives us this example:

> *"A change in nucleotide sequence of the gene's coding region may lead to a different amino acid being added to the growing polypeptide chain, causing a change in protein structure and function. In sickle cell anemia, the hemoglobin β chain has a single amino acid substitution, causing a change in protein structure and function. Specifically, the amino acid glutamic acid is substituted by valine in the β chain. What is most remarkable to consider is that a hemoglobin molecule is made up of two alpha chains and two beta chains that each consist of about 150 amino acids. The molecule, therefore, has about 600 amino acids. The structural difference between a normal hemoglobin molecule and a sickle cell molecule, which dramatically decreases life expectancy—is a single amino acid of the 600. What is even more remarkable is that those 600 amino acids are encoded by three nucleotides each, and the mutation is caused by a single base*

change (point mutation), 1 in 1800 bases." [7]

This functionality of DNA being translated to RNA and then used to build proteins is fundamental to present-day Biology, as we can see from this quote.

"In 1958, five years after he helped discover the double helix structure of DNA, Francis Crick coined the term 'Central Dogma' to characterize the all-important cellular processes whereby DNA is 'transcribed' into RNA and RNA is 'translated' into protein." [41]

Francis Crick, along with James Watson, was one of the people credited with the discovery of the structure of DNA. Francis Crick chose the word "dogma" to describe the process by which information is transmitted from DNA to RNA then to the proteins that are the building blocks of the distinct cells that make up organisms. The image below diagrams the central dogma of Biology.

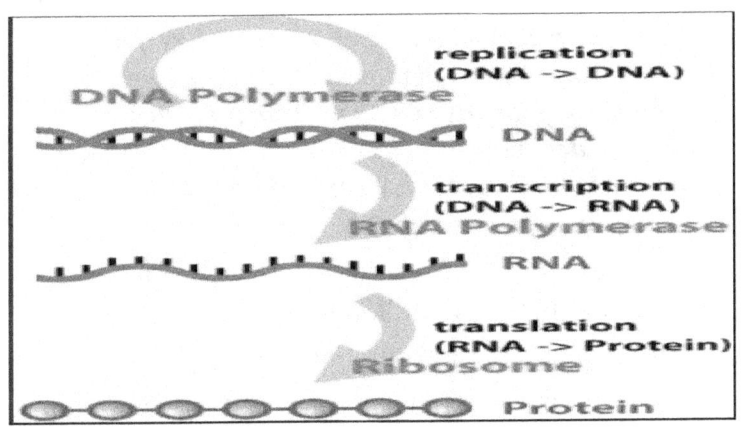

Central Dogma [7]

Notice that the central dogma refers to a one-way flow of information from DNA to proteins, and this quote emphasizes that.

> *"Since then, much effort has been devoted to constructing a flow chart for the various forms of molecular information. One of the fruits of this labor has been a sense that many of the main information highways are known —for example, the central dogma that describes the irreversible flow of information from DNA to RNA to proteins."* [35]

Proteins are what makes organisms function. In that sense, we are made up of proteins. The following citation states that we are made of proteins and again emphasizes the one-way direction of the flow of information.

> *The central dogma is often expressed as the following: "DNA makes RNA, RNA makes proteins, proteins make us." Protein is never translated back to RNA or DNA. Furthermore, DNA is never translated directly to protein.* 14

Note that in the central dogma and the previous diagram, the flow of information is from DNA to RNA to proteins. There is no information flowing in the opposite direction.

Coding DNA contains all the information needed to make more than 20,000 different proteins. That is a massive amount of information. As we saw, in the case of hemoglobin, it requires 1,800 nucleotides to indicate in what order 600 amino acids are joined to make hemoglobin proteins. As we saw, even one mistake can cause defective protein. So DNA contains precise information on how to build proteins. It is correct then to look at DNA as a container of information.

At this point, we have established that DNA contains a massive amount of information that is essential to build an organism. We have also seen that this is precise information. In many cases, if a single nucleotide varies we lose the needed functionality. Information flows from this container out to the organism. The structure of every protein must be specified in DNA nucleotides. *It is not yet known how most of the information, needed to define structures larger than proteins is encoded.* Everyone assumes this information is in DNA. This structural information will also be encoded in nucleotides, and the flow of this information will also be from nucleotides to the organism.

Now that we are acquainted with how precise and detailed this information is, we can appreciate the question raised in this book. Where does the information in DNA come from?

Chapter 4
Abstract Information

The complexity of the information encoded in DNA is even more amazing when we see that it is in an abstract form.

When we speak of information, such as that in DNA, we can be more precise and refer to it as abstract information. I use the word abstract as it means, **"*disassociated from any specific instance*"** *or* **"*expressing a quality apart from an object*"** [26]

Why do I say the information in DNA abstract? There are no amino acids in DNA. Nor are there any *peptide* chains or proteins in DNA. There are some 20,000 different proteins, 22 distinct amino acids, and only four types of nucleotides. Information in DNA is in the form of an abstract language.

Let me give you a simple example of abstract information.

Suppose I had shared some delicious brownies my daughter made with a friend. A few weeks later, this friend asks if I can give him the recipe for these brownies and would like it right away. So I decide to send a text with our family's recipe 99 for brownies.

As I begin to key in the recipe, I see all the characters in the alphabet displayed on my screen. I touch on the characters in the desired sequence to indicate each letter or digit of the recipe. The software in the phone takes note of the x-y coordinate that I touch on the screen and knows what character is displayed at that location. The software translates this x-y coordinate into its corresponding character. Our smart-phones are digital computers. Digital computers store all their information in binary format. In binary code, there are only ones and zeros. All the letters, digits, or punctuation marks we use are encoded into the ones and zeros of binary in sequences determined by an accepted character conversion table. So this character gets converted to its binary value assigned by some character conversion table. This binary value gets stored in some temporary electronic memory on the phone.

Once I have finished keying in the recipe, I click send. The message is converted to variations of radio frequencies that are broadcast. Somewhere there is a cellular tower that receives this radio signal and converts it back to its binary value in some character set. This binary information is then transmitted over the digital network of the phone carrier as close as it can get to the specified destination. If the person receiving the recipe is also on a cell phone, the whole process is reversed. The message arrives in binary format to the nearest cell phone tower, where it is converted to variations of radio frequencies that are broadcast and picked up by the receiver's phone. Once the characters are received on their phone, they are displayed as a set of small dots that draw the character on the screen. The person receiving the recipe can then get the ingredients specified, follow the instructions given, and hopefully produce some brownies with the flavor and texture close to our family's favorite.

If one of the people tasting the brownies likewise asks my friend for the recipe, then my friend might offer to print it out for them. Thus the same abstract information would be transmitted in yet another form.

Until the person begins to acquire and assemble the ingredients, there is no chocolate, sugar, or flour. There is no flavor, no smell, or anything you would find in brownies. There is only abstract information about brownies. Undoubtedly there is value in discussing the touch screen involved, the character sets used, the radio frequencies utilized, the cell towers, and all the technology involved. In the case of the printed copy, we could talk about the technology used to transmit the text to the printer or the printing technology used to print the recipe. We could even talk about the type of paper on which the recipe was printed. But what is most important is the information itself. It is brownies, the ingredients and the steps to make them. The medium, mode of storing, and transporting this information are secondary to the origin of the information and the functionality contained in the information.

In the same way, whether genetic information is in the form of nucleotides in the DNA chain, transposed into bases in RNA, or even a sequence of amino acids in a polypeptide chain, it is the information that determines the makeup of the protein. The functionality of the encoded protein is what is important for an organism. The information is of fundamental importance.

Let me underline then, the abstract nature of the relationship between DNA and amino acids and proteins. The information in DNA, that causes the organism to have feathers and be a bird, or to have a shell and be a turtle will also be abstract. And it is not yet understood.

> *"Only about 1 percent of DNA is made up of protein-coding genes; the other 99 percent is noncoding. Noncoding DNA does not provide instructions for making proteins. Scientists once thought noncoding DNA was "junk," with no known purpose. However, it is becoming clear that at least some of it is*

integral to the function of cells, particularly the control of gene activity." 33

For DNA to determine the structures of organisms, it must have a way to specify how to build an organism.

We know in coding DNA, that a codon, or three nucleotides, are needed specify an amino acid. Consider how many nucleotides will be required to specify more complex structures of organisms. Substantially longer sequences of nucleotides will be required to specify these complex structures. For instance, how many nucleotides will be needed to specify the curvature of the hip bone socket? Or, to indicate the position of nerves and blood vessels in a hip bone socket?

It will be much more challenging to decipher the more complex patterns in the nucleotides that specify structures used only once in an organism. In advance, we know that the DNA will be an abstract representation of the structures defined.

As explained in biology's central dogma, the origin of all proteins in organisms has to be traced to the origin of information in DNA. In the same fashion, any evolution of proteins in organisms has to be traced to the origin and change of information in DNA. As we've seen, when we speak of information in DNA, we are talking about specific nucleotides and their order within DNA.

When we look at a field of fossils, we are looking at the composite results of sequences of nucleotides in DNA. When we look at known organisms organized on a tree of life, we are looking at the end product of sequences of DNA nucleotides. *Fossils and trees of life help us theorize about the origin and evolution of organisms. But the final explanation has to be in the origin and evolution of the information contained in data sequences of nucleotides in DNA.*

The central dogma tells us that there is no flow of information from an organism to proteins to RNA to the nucleotides in DNA. We will proceed to examine other interactions that can and do change the nucleotides in DNA. Some of these are proposed to explain the information in DNA. We will see if any bring information into DNA.

So any change or evolution in the structures of organisms has to occur in the abstract language or code of DNA. The abstract nature of this information represents a barrier to any flow of information from the organism, the environment, or any other interaction to the DNA sequence. The only way there could be a flow of information would be to translate the information, from its description of functionality to a sequence of DNA nucleotides. Further ahead, we will work through a test case to understand this barrier.

So what is it that changes the information in DNA? We're going to look at that question in Part 2

PART 2

The Source of Information

Chapter 5
Intelligent Agents Can Add or Change information in DNA

In Part 1, we established that DNA is the container that stores the information necessary for life to originate in a single cell that multiplies until it becomes a complex organism. The information stored in DNA is abstract information, that is, information that is independent of the object that stores it. Any changes to the information DNA contains must be made in this same abstract language of DNA.

Now we turn our attention to the question we were left with: how can information be added to DNA? In subsequent chapters, we will examine whether the standard explanations offered by scientists—mutations, the environment, and natural selection—are plausible. But before we do, it is informative to look at a case where we know information can be added to DNA. It will provide a sharp contrast when we look at the three other cases.

Recent advances in genetics have opened up a new way to add or change the information in DNA. I'm referring to gene-splicing, gene-editing, or *recombinant DNA,* which are all done by humans. With the knowledge of how proteins are encoded coupled with an understanding of how proteins affect or determine function in an organism, scientists can think of ways to improve organisms. They are also cases where they use organisms as tools for their work on DNA or to mass-produce specific proteins.

In these cases, scientists suggest new or better information and edit DNA sequences to add that information. They translate what they desire in the organism to the language of DNA. This design and translation are a man-made way to overcome the one-way flow of information from DNA sequences to the organism, as expressed in the central dogma.

We will briefly look at three examples of using gene editing to improve organisms.

Synthetic Human Insulin

Synthetic human insulin was the first golden molecule of the biotech industry and the direct result of recombinant DNA technology. Currently, millions of diabetics worldwide use synthetic insulin to regulate their blood sugar levels. Synthetic insulin is made in both bacteria and yeast. 12

We all know someone who has to take insulin to control their diabetes. What you may not know is that bacteria or yeast "factories" manufacture a lot of this insulin. Scientists extract from human DNA, the nucleotide sequence of the gene that produces insulin. They then insert this nucleotide sequence either into a bacteria or a yeast's DNA. When this bacteria divides, it has a copy of this human insulin gene. As this recombinant bacteria produce proteins, among them is the "synthetic" human insulin. This process is more complicated than what I have explained here but very effective. Thousands of people benefit from this synthetic insulin.

This illustration shows how the nucleotide sequence of the insulin gene is inserted into the DNA of bacteria, which, as you can see, is in a circle rather than strands.

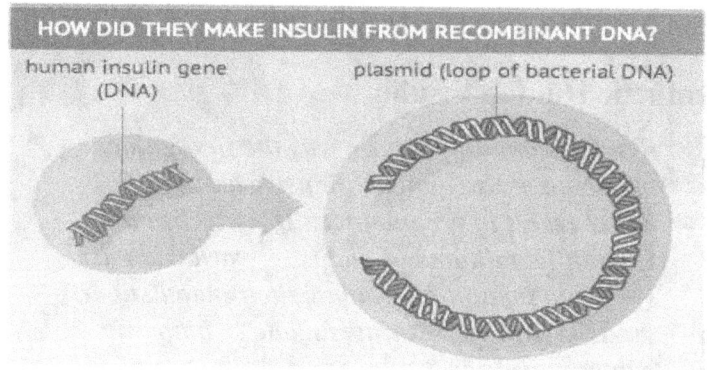

Synthetic Insulin (8)

Synthetic insulin was first made back in 1978 by Genentech, Inc., and City of Hope National Medical Center. 20 So this technology has been around for several decades.

Bt Cotton

Cotton is an excellent example of genetic engineering. Cotton is a beneficial plant that is produced and consumed by people all over the world. Not only is cotton used for its fiber, also cottonseed oil is used as a base ingredient for many foods we consume. Cotton seeds are also used as animal feed and to make certain plastics and even medical supplies. 15

Insects, such as the boll weevil, are natural enemies of cotton. The boll weevil feeds on both the buds as well as the flowers of cotton. More than one hundred years ago, a Japanese biologist discovered a bacteria called Bacillus thuringiensis (Bt). He saw that this bacteria caused disease in silkworms. Later, other researchers saw that it would kill different insects. A few decades after its discovery, farmers began to use it as an insecticide, which they sprayed on their crops. Continued research on Bacillus thuringiensis found that a particular group of proteins called Cry proteins were the toxins that killed insects. 36

With the advent of genetic engineering, scientists decided to add the DNA sequences that encode for Cry proteins into the DNA of corn and, later on, cotton. This addition gave these plants the same insecticide qualities. These Cry proteins in this modified cotton protect it from the boll weevil.

This genetically modified cotton is known as Bt cotton. It was recognized as an invention of Monsanto, and they were given a patent for it as their intellectual property. This genetically modified cotton was such a good idea that it represents more than 80% of cotton grown in the United States, and at least half the cotton grown in China. 38 Bt cotton has been used for more than 20 years in Mexico. Bt cotton has reduced the various costs associated with spraying cotton with pesticides. 42

Loblolly Pines

Forestry companies farm trees for use as lumber, paper, wood pulp, and as an energy source. Some of these companies have also used genetic engineering to develop trees more fitted to their needs.

Arborgen, Inc. is a company that has genetically engineered a better tree. It is reported that they inserted genes from several other organisms to make their AGV Varietal Loblolly Pines. They inserted genes from the Monterey Pine, American sweetgum trees, and others. They report several benefits in their genetically modified Loblolly pine compared to non-modified Loblolly pines. One significant advantage is that the genetically modified tree has a higher wood density. Higher wood density means the wood is stronger and more durable. Higher density also means it has higher energy content if the purpose is to use the tree for fuel. 22

An additional benefit is that you can harvest the first thinning in as few as 8 to 10 years. They claim this first harvest will contain up to 50% more solid wood. When the trees are harvested they should provide some 80% more sawtimber . 2

Now let's look at what these three cases of genetic editing have in common. First of all, these changes to DNA could not have occurred just by selective breeding. Selective breeding would seek to favor the organisms which already have certain genetic traits to obtain a breed where that trait is dominant. In these cases, the DNA sequences of the nucleotides A, C, G, and T, were not in the original organisms. It was not enough to change one nucleotide here and another there. Complete protein or polypeptide DNA sequences were inserted into the DNA of the target organism.

The second thing we notice is that the external agent modifying the information in DNA is an intelligent agent. That is to say, the foreign DNA sequences inserted into the organism were deliberately chosen because they were known to contain information that made a different organism function a certain way. Therefore it was hoped that the target organism would change to function in that way. The intelligent agent worked backward, taking the desired end-state of the organism's functionality and then injecting that new information into the DNA. It was not done gradually or piecemeal. Instead, the complete protein sequence was introduced into the DNA.

An intelligent, external agent can change DNA. So much so that patents have been granted for these genetic modification inventions to protect the intellectual property of the intelligent agents who made these changes to DNA.

Chapter 6
The Evidence is in the Lab

We have seen that an intelligent, external agent can add information to DNA. The question remains, can information be added to DNA through natural processes? Before we address this question, we will first briefly review the standard by which we will judge whether this can be done, the scientific method.

Roger Bacon (1214–1294) is generally credited as having proposed the scientific method. The scientific method consists of observation, induction, hypothesis, test by experiment to arrive at knowledge. In his Opus Maius he wrote:

> *Having laid down fundamental principles of the wisdom of the Latins so far as they are found in language, mathematics, and optics, I now wish to unfold the principles of experimental science, since without experiment nothing can be sufficiently known. There are two ways of acquiring knowledge, one through reason, the other by experiment. Argument reaches a conclusion and compels us to admit it, but it neither makes us certain nor so annihilates doubt that the mind rests calm in the intuition of truth, unless it finds this certitude by way of experience.* 4

Bacon's sequence was very logical. Observation consists of looking carefully at what you want to understand. Induction is thinking about it, reasoning through it. A hypothesis is creatively coming up with a possible explanation of what you have observed. Test by an experiment consists of thinking of some conditions and processes which could give you evidence of the validity or invalidity of your hypothesis. Based on the results of the experiment, you gain knowledge, be it evidence of the validity or invalidity of your hypothesis. The worst-case scenario is you get evidence of your experiment being poorly thought out or designed!

Bacon pointed out that through an experiment, you experience the results. Expressed another way, "experimental" science gives evidence that a hypothesis is valid or invalid. Bacon realized the limitations and prejudices of our minds and our context. For this reason, he felt it was essential to publish the details of the experiment so others could duplicate the same experiment. If others obtained the same results of your experiment, then your hypothesis begins to be shown to be valid beyond personal inclinations or context. The results should be independently verified, and you should document the results so that others might repeat the experiment.

Popular media and common folk still talk of proof and disproof, but today scientists speak of probability or degrees of certainty, rather than simply just proof.

We can show how Bacon's method works with a simple example. Pure water, H2O, at standard sea-level pressure, boils at 100 degrees centigrade. If you doubt that statement, you can set up an experiment of your own, with pure water, H_2O, at standard sea-level atmospheric pressure at 100 degrees centigrade and see if it boils.

In this case, you could be a capitalist or a communist. You could be of any race, you could be on land, in outer space, or under the ocean. You could believe that life has a purpose or not. That does not matter. You can set up this experiment, with pure water, H_2O, at atmospheric pressure equal to sea level and the water at 100 degrees centigrade it will boil. If it does not boil, you should double-check the conditions specified. I can say this confidently because the experiment has been repeated countless times, and it always yields the same result.

In the last chapter, we looked at three cases of gene editing, synthetic human insulin, Bt Cotton, and AGV varietal Loblolly Pines. These are examples of reproducible experiments that give validity to our statement that DNA contains abstract information that builds organisms. As cited previously, that "DNA sequences provide the basis for the genetic code for all of the proteins synthesized by our bodies." These verifiable instances of genetic engineering give evidence that by moving DNA sequences from one organism to another, the receptor organisms synthesized new proteins. With these new proteins, characteristics of one organism moved to another organism.

There are dozens of similar experiments that produce verifiable and reproducible results that confirm that DNA contains information that makes proteins and influences the characteristics of organisms. Manipulating DNA in this fashion is no longer considered experimental, in the sense of waiting to see what happens because the outcome is uncertain. We could say there is functional certitude of results, meaning that the organism can reliably be expected to function in the new intended way when the external DNA sequence is inserted. Moving a section of DNA base pairs to change an organism's characteristics or to move characteristics from one organism to another is no longer something hypothetical, it is something scientists can do, and in fact, do.

All this work in genetic or DNA modification gives us functional certitude that DNA contains information that is used to build organisms.

These advances in genetics are awe-inspiring and beneficial. But they fall short of proving that *all* the information required to make an organism is stored in DNA. As stated earlier, there is a whole lot we still do not know about how the information to build complex structures, made up of multiple types of proteins, is stored in DNA.

The idea that *all* the information needed to build an organism is in DNA is a presupposition that is yet to be shown to be valid by experiments. Any claim about the content of information stored in DNA, or how it can be changed, must pass Bacon's test to be believable.

Chapter 7
What Else Changes DNA Information?

When we looked at gene editing in Chapter 5, we saw that an external, intelligent agent can modify the information contained in DNA to the point that the functions of an organism can be altered and even introduce new functions. Now we turn our attention to the far more interesting question of whether it's possible for a non-intelligent process such as mutations to improve the information DNA contains or add to it. By non-intelligent, we mean that, unlike gene editing, there is no intent or end-goal in mind. Gene editing starts with the desired functionality and works backward to make the necessary changes to DNA. Can natural changes to DNA lead to changes in the functionality of an organism?

Recall from Chapter 3 that DNA information is stored in nucleotides. So to be more specific, we are asking these three questions. What changes the nucleotides in DNA? What adds or takes away nucleotides? What changes the order of nucleotides?

Changes to the nucleotides or nucleotide sequences are called mutations. There are two categories of causes of mutations. If a mutation occurs in the organism's normal process of replication or making copies of DNA, we would call this a spontaneous mutation. If some influence outside this normal replication causes a mutation, it would be considered an induced mutation.

Let's talk about spontaneous mutations first. Every cell in your body has a complete set of DNA. A copy of the cell's DNA is made when a cell divides into two cells. With this copy, each of the new cells gets a complete set of DNA. Making this copy of DNA is known as DNA replication. In this process, simple copying errors or typos can occur. There are mechanisms and procedures to correct or reject copies with mistakes in them. Any error that makes it through this correction process to the final copy is a mutation, in this case, a spontaneous mutation.

These errors, typos, or *polymorphisms* can happen at any "letter" or nucleotide. The most common mutations only affect one nucleotide, [34] but a mutation could affect more nucleotides.

Errors made while copying DNA sequences are not made with the intent of adding or improving information in the DNA nucleotide sequence. Nor are these mutations made to harm the information in DNA. Mutations happen without any intent at all. These changes occur as failures of the replication process. Not only do they happen with no intent, there is also no particular place in the DNA sequence that they occur.

There are five major types of mutations that we will briefly examine: missense, silent, nonsense, deletion, and insertion mutations. These types of mutations may occur as a result of either spontaneous or induced mutations. We will consider the effects of these mutations on coding DNA.

Missense Mutation

This type of mutation would change one nucleotide in DNA for another. In the following example, the change is within a sequence of coding DNA. By changing one nucleotide, the amino acid gets changed and probably changes the functionality of the protein.

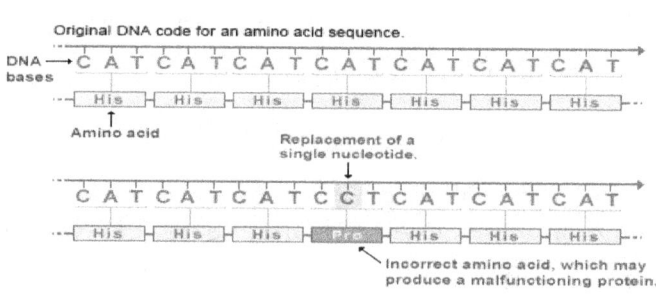

Missense Mutation (9)

Silent Mutation

When a single nucleotide changes and that does not change the amino acid that is encoded in the codon, then it is called a silent mutation. A silent mutation would not change the protein at all.

Nonsense Mutation

When a single nucleotide changes and it converts the codon to a stop codon, then it is called a nonsense mutation. A nonsense mutation would cause the protein to be non-functional.

Deletion Mutation

Another mutation is where one or more nucleotides are deleted. Here is an example of deleting one nucleotide from DNA.

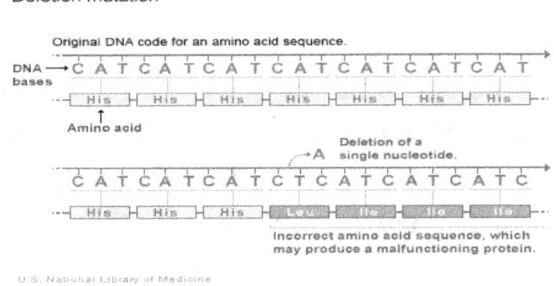

Deletion mutation (10)

As you can see in the illustration, if only one nucleotide is deleted, then all the following nucleotides shift by one nucleotide. Groups of three nucleotides form a codon and determine the amino acid to add next to the protein. Thus a single deletion affects the codon the deletion is in, plus all following codons to code for different amino acids. This effect is called a reading frameshift.

In the case of protein-coding DNA, if we shift one or two nucleotides due to a deletion or an insertion, we're going to change many of the amino acids and, thus, the functionality of the protein.

The illustrations we present focus on the effect of mutations on coding DNA. Coding DNA is DNA that codes for proteins, or makes proteins. Only 2% of DNA codes for proteins. Mutations can happen at any place in the DNA nucleotide sequence. The complete DNA sequence is copied when a cell replicates its DNA as it divides into two cells. The 90% or more of DNA that we do not know what it may be used for is also copied. In one of the later chapters of this book, we will consider changes to DNA that affect non-coding DNA. We will to understand the possibilities, dangerous, and otherwise, for changes in DNA.

Insertion Mutation

Another type of spontaneous mutation would insert one or more nucleotides in DNA. Here, we have an illustration of an insertion mutation. Similar to the case of a deletion mutation, if a single nucleotide is added to coding DNA, all the other nucleotides will move down by one nucleotide and change most of the amino acids specified from thereon.

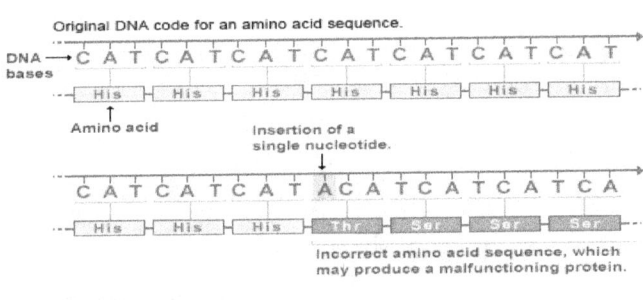

Insertion Mutation (11)

These are simple explanations of these mutations and their effects. You can begin to see the wide-ranging effects of mutations.

Another spontaneous mutation would *translocate* a group of nucleotides in DNA. Translocate is to take a group of nucleotides and move it from one place to the other. There are some genes that do this as part of their normal behavior.

Another mutation would invert a group of nucleotides. When DNA is duplicated, pieces of DNA on one side are duplicated backward and flipped around. For whatever reason, sometimes, a sequence of nucleotides in DNA gets inverted.

When DNA is transcribed into RNA, errors can also be made. These errors cause changes in the sequence of bases in the proteins to be translated from the RNA. They do not change the DNA sequence from which the RNA was copied.

Induced Mutations

Unliked spontaneous mutations, induced mutations don't occur during replication but are caused by environmental factors called *mutagens*.

Diverse types of radiation and chemicals can cause changes to the nucleotides in DNA. Some examples of these mutagens would be sunlight, cigarette smoke, and radiation, which are known causes of changes to our DNA. These mutagens come from the external environment.

Here, we have an illustration of a mutation caused by incoming UV photons.

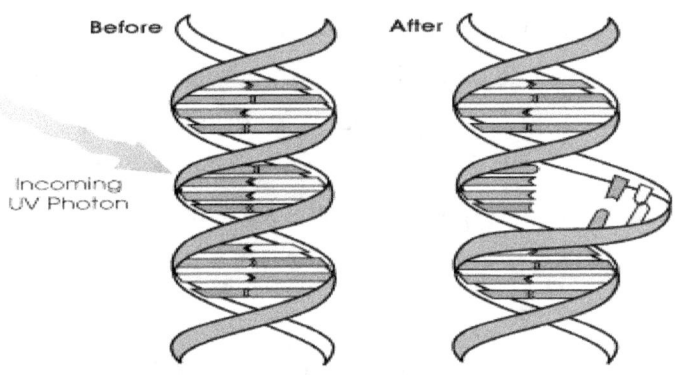

DNA Mutation Caused by UV Light [12]

One of the tools of geneticists is to induce mutations to try to learn about the function of genes. Some of the scientists wanting to study genes and the code of DNA subjected DNA to x-rays and this exposure to radiation produced some mutations. They did this so they can investigate the effects of those different mutations.

Induced mutations, like spontaneous mutations, do not occur with the intent of sending information to DNA or destroying information in DNA. The environmental mutagens occur with no intent at all. Any part of the DNA sequence can be affected by environmental mutagens. These mutagens have no rhyme or reason in their effects.

Viruses and Bacteria

There's a different type of change to DNA. I'm not

sure it falls in either those two categories. This modification is caused by the interchange of DNA data or nucleotide sequences between organisms of different types. The classic example of this is bacteria. Bacteria of one type have been shown to exchange DNA data or nucleotide sequences with bacteria of another kind.

Viruses depend on host organisms to reproduce their DNA or RNA. Here we can see some illustrations of those scary viruses.

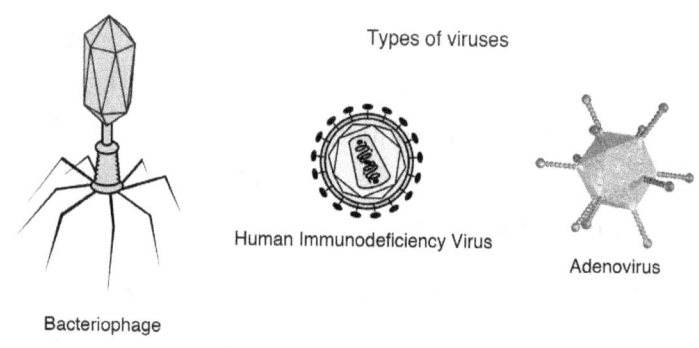

Types of Virus (13)

Germline Mutations

As I have pointed out in previous chapters, these changes occur in the abstract nucleotides of DNA and the abstract information they contain. Nucleotides are just like letters in an alphabet. Words are made from these letters. And sentences are formed from those words. In the case of coding DNA, the nucleotides make amino acids, and amino acids in their sequence make proteins. So, these changes occur in the abstract nucleotides, a similar situation, as having some text and we start kicking around some letters, adding letters, deleting letters, swapping letters, inverting letters, not far from that, when we talk about changes in DNA.

Most mutations happen in DNA in the normal cells of our body and not in our germ cells, that is, sperm or egg cells. We refer to these as *somatic* mutations. One well-known example you are probably familiar with is that exposure to sunlight can cause mutations in our skin cells that can lead to skin cancer. These kinds of mutations, somatic, won't be passed on to the next generation.

A male's sperm or a female's egg are called germ cells. DNA is copied by a unique process for a male's sperm or a female's egg. We have two copies of each of the 22 chromosomes in our DNA these are blended to make a new unique copy for each germ cell. We also have two sex chromosomes. Germ cells will only have one copy of each chromosome.

Mutations that occur in the DNA of our germ cells are called *germline* mutations. Germline mutations are passed on to children, from one generation to the next.

If an offspring inherits a germline mutation from their parents, every cell in their body will have this change in their DNA.

Somatic mutations do not change the information repository of the subsequent generations, so they cannot explain the repository of information in DNA. Germline mutations do change the information repository of subsequent generations, so they could add to or take away from the information stored in DNA.

We could take years studying this. This chapter gives us a quick introduction to the types of mutations. As we dive deeper into the question of how the information we see in DNA got there, we want to see whether mutations could be responsible for accumulating that information. We have set the conditions by which those changes if they did occur, would have to happen. For starters, the mutations would have to add the information to the germ cells. The mutation would have to add abstract information to the information contained in the nucleotides of the germ cells in one of the ways we described in this chapter.

Chapter 8
Relationship Between Mutations and Information in DNA

There's a lot of misunderstanding about mutations and the information in DNA. There are three questions we can ask to clear up misconceptions about changes in DNA.

1) What is the relationship between these mutations and the information in DNA? We've shown in previous chapters that DNA contains information. What is the correlation, if any, between these mutations and the information in DNA?

2) What is the relationship between these mutations and the structure of organisms built or based on DNA information?

3) What is the relationship between these changes and the environment in which the organism lives?

You need to use several adjectives to explain adequately the answer to all three questions. All three questions have the same answer, the same explanation. The reason I use different adjectives is, sometimes, we get hung up, we think it's just a word, and we think we're going to argue a word, it's not a word, it's a concept. And the concept, being somewhat broad, can be explained best by several adjectives. These would be the adjectives that I would want to use.

Random: *Definition:* Happening or done, by chance rather than according to a conscious decision, a plan, or pattern. *Application:* This is the first adjective used by geneticists and even evolutionists. Neither spontaneous nor induced mutations occur according to a conscious decision, plan, or pattern. For this reason the word random is a fitting description of mutations. Mutations are not generated with the purpose of being packets of information sent to DNA. These changes are independent of information in DNA, independent of the structure of the organism, and independent of the organism's environment.

In IT we speak of data noise, or noisy data. *"**Noisy data is data that is corrupted, or distorted.**"* [54] Mutations can be understood as noise that will be added into our data stored in nucleotides. These changes are also random since we do not know what effect they may have on the information encoded into the nucleotides.

These mutations may be fatal, they may be detrimental, they may have no effect, or they may be helpful. A mutation would be fatal if it makes the existing information non-functional. A mutation would be detrimental if it changes the information so that it builds a less functional protein or breaks other functionality determined by a specific sequence of DNA nucleotides. If a mutation has no effect, as in the case of a nonsense mutation, then the mutation neither added nor took away from the functional information in DNA. If the mutation is fortuitous, then it would be helpful, and in that case the mutation improves upon the functionality of the protein or other functionality. Only in this case does a mutation improve the information in DNA.

The changes are not random in the sense of guaranteeing that they will be different each time, or that, eventually, they change every nucleotide of DNA. They are random in the sense that their occurrence is independent of any intent or desired change in the functionality of an organism.

Unrelated: *Definition:* Not connected by reason of an established or discoverable relation. *Application:* These changes are not connected by reason of an established or discoverable relation. The cause of a spontaneous or induced mutation has no relationship with the effect it might have on the information in DNA. Spontaneous or induced mutations have no relationship to the location on the DNA sequence that they change. So, the changes are unrelated to information in DNA. The existing DNA sequence produces the structure of the organism. Since the mutations have no relationship to the location of DNA they affect they're also unrelated to the structure of the organism. The mutations have no relationship to how the organism survives in its environment.

Irrelevant: *Definition:* Not connected with or relevant to something, not applicable to something. *Application:* The causes and destination of mutations are not connected with the information in DNA. They have no application to the information in DNA. The changes have no relevance to the structure of the organism. In the same way, they're not applicable to the organism's environment.

Accidental: *Definition:* Happening by chance, unintentionally, or unexpectedly. Occurring without intent or through carelessness, often with unfortunate results. *Application:* Another adjective would be accidental. The mutations would be accidental as they arise from extrinsic causes. They are accidental as they occur unexpectedly or by chance. Lastly, they are accidental since they happen without intent and, often, with unfortunate results. They are failures to make an exact copy of DNA. Or they are environmental mutagens that just happened to interact with DNA. And, very often, they have unfortunate results.

Incidental: *Definition:* Occurring merely by chance, or without intention or calculation. Causing only minor consequences. *Application:* The word incidental is similar to accidental. The difference is the implied results. Accidental often implies unfortunate results, whereas incidental implies minor consequences. They occur merely by chance or without intention or calculation and they have minor consequences. No one proposes that these changes are made intelligently or with some intention or with any calculation.

Fortuitous: *Definition:* Occurring by chance, but also fortunate or lucky. Produces a good outcome. *Application:* We have explained that mutations occur by chance. It's rare that a mutation is going to be fortunate or lucky. But sometimes, the changes are by lucky chance. They're fortuitous, and they have a good outcome. Fortuitous mutations change the information in functioning DNA and improve the functionality of a protein or other functionality in DNA.

We argue that these adjectives describe the effects of mutations. As we have explained, mutations are not generated as information packets to send into DNA. They are random in origin and independent of the organism. While in rare fortuitous cases a mutation can improve the information in functional DNA, they are correctly understood as a source or random data noise. The question remains, does this random noise affecting DNA sequences explain the origin of information in DNA? We will examine this possibility in Chapter 12.

Chapter 9
Does Environment Change DNA?

When we ask if the environment can change DNA, what we mean is, does the environment change the nucleotides in DNA? The simple answer is yes.

We had spoken about changes in the nucleotides of DNA as being mutations. And there are two types, which were spontaneous and induced mutations. When we talked about induced mutations, we explained that environmental mutagens cause induced mutations. Thus the environment changes the nucleotides in DNA. But it is necessary to understand how the environment changes DNA and how it does not change DNA.

These agents are called environmental mutagens because they are in the environment. Diverse types of radiation and chemicals can cause changes to the nucleotides in DNA. Previously we showed an illustration of radiation changing some nucleotides. This radiation could come from the sun or some other source, and it can change DNA sequences.

While these modifications come from the environment, they do not transmit or bring any information about the environment in which the organism lives to the DNA. I used several adjectives to explain how these environmental mutagens change DNA: random, unrelated, irrelevant, accidental, incidental, and (possibly) fortuitous.

The changes caused by environmental mutagens are independent of the information in DNA. They are also independent of the structure of the organism, or the complete functioning organism. And the changes caused by environmental mutagens are unrelated to all other conditions in the environment. Let's look at an example to clarify these three statements.

Solar flares send gamma rays to planet earth. The gamma particles in gamma rays can cause changes in DNA and, thus, mutations in an organism. Solar flares are caused by changes in magnetic field lines near sunspots on the sun. These solar flares happen due to interactions on the sun and have no relationship at all to planet earth, much less to any particular organism on earth. There is no reason for them to interact with DNA and the information encoded in the nucleotides. The organism, with its DNA just happened to be in the path of these gamma particles. These gamma rays and particles can cause mutations that affect bones and any other part of an organism. But again, the gamma rays are going to be sent, and have no relation to whether an organism could use some changes. The gamma rays are going to come whether there is an ice age or global warming. This is what is meant by saying they are independent of the information in DNA, the structure of an organism, and the conditions of the environment.

Geological events like the eruption of a volcano, or emissions from thermal springs could add mutagens to the water or air. Physical interactions cause these events. They do not occur with any intent to affect living organisms. If they affect the germ cell of any organism, it is accidental. The organism just happened to live in the area affected by these mutagens. These mutagens could affect any position in the DNA chain, like throwing darts while blindfolded, they are not aimed at any particular target. Some of these geological events could affect the environment. But their mutagen effects on organisms have no relation to where there is abundant food supply in the environment or any other factor favorable to an organism.

So these mutational agents come from the environment. But they do not change DNA such that the functional and structural characteristics of the organism, the *phenotype,* is more fit for conditions in that specific environment. Biologist Chrissy Spencer explains this further:

Mutations are not caused or induced as a

result of environmental change. Variation is already present in the population. When the environment changes, those individuals that already have some beneficial variation (mutations) that is well suited to the new environment are more likely to survive and reproduce; organisms do not develop new mutations in response to the environmental change. (And if there is no variation present in the population such that some individuals survive and reproduce, then the population is likely to go extinct). 47

Environmental mutagens do not bring information from the environment into DNA.

This has been a simple chapter. We just wanted to be precise in our answer. Yes, the environment changes DNA in one sense of the term, as random, unrelated, undirected noise that may modify the nucleotide data pool. We have adequately explained this change in this chapter.

In the next chapter, we will examine the proposition that the environment could change DNA as if it were sending information about the environment to DNA. We will show why this is not possible.

Chapter 10
Does The Environment Send Information To DNA?

Jean-Baptiste Lamarck promoted an idea called inheritance of acquired characters before Mendel's studies of the basic principles of *heredity,* Darwin's theory of natural selection, and more than a century before the discovery of DNA. This idea states that there is a relationship between the environment and inherited changes. The following citation is a summary of what Lamarck believed.

> *If an organism changes during life in order to adapt to its environment, those changes are passed on to its offspring. He said that change is made by what the organisms want or need.* 5

Since Lamarck wrote before Mendel, he did not know how inheritance works. He did not offer any proof of inheritance of acquired characters. ***He believed it was so transparently obvious that it needed no assemblage of facts or trial by experiment to confirm it.*** 8 Lamarck and others of his time were making assumptions regarding heredity, of which they didn't have even basic knowledge.

It did not take long to show that inheritance of acquired characters was not a valid hypothesis.

Lamarck (1744-1829) proposed that

> *characteristics acquired by an animal in its life could be inherited by its offspring. For example, generations of primitive, short-necked giraffes that stretched their necks to reach high branches all their lives could pass down this character that they had developed for slightly longer necks. This ultimately resulted in today's long-necked giraffes - according to Lamarck.*
> *But experiments in the 1890s (cutting off 20 generations of lab rat tails) showed that changes in the body couldn't change inheritance. Also, shepherds have docked the tails of lambs for thousands of years. Nevertheless, lambs are born every year with tails intact. - DiVenere, V* 18

Fast-forward a couple of centuries, and people are once again affirming that the environment changes an organism so that it can better adapt to its environment and these changes are passed on to their offspring. They propose all of this within the framework that organisms live within an environment, are subject to random mutations, and undergo natural selection.

The argument is as follows: in a given environment, if a mutation helps a particular organism reproduce more offspring than other organisms of its same kind, and if these offspring are also able to grow and reproduce more, then these mutated offspring are favored by natural selection. For the sake of brevity, I will refer to this as *effective reproducibility*. A change in the environment can affect any organism's effective reproduction rate. This change not only affects organisms with new mutations. Organisms without a new mutation may also see increased or decreased effective reproduction due to this change in the environment.

The question is whether the environment changes DNA, in a way that would send information from the environment into DNA. Recall that we have defined changing DNA as changing the nucleotides in DNA.

The Question of Bipedalism

Let's look at one example, and examine whether the environment can change DNA. Many proposals have been offered to explain the development of upright walking or bipedalism in humans. Why has there been so much interest in this particular functionality? Here is a short explanation of the importance of bipedalism.

> ***Walking upright on two legs is the trait that defines the hominid lineage: Bipedalism separated the first hominids from the rest of the four-legged apes. It took a while for anthropologists to realize this.*** 53

Most of the proposals regarding the origin of

upright walking are tied to changes in the environment. You have probably heard a popular scientific explanation like this.

> **"Upright-walking was a response to environmental changes in East Africa at the time. The rainforest was turning into steppes and grasslands due to global cooling. This reduced the apes' habitat and drove two-thirds of their species into extinction. Some primates developed arboreal lifestyles and became the gibbon lineage. Others developed an upright stance as a survival characteristic, to see over tall grasses and spent time more on the ground."** James Schombert - 21st Century Science - University of Oregon 44

Prof. Schombert's statement that upright-walking was a response to environmental changes is in accord with the savanna hypothesis, which is still widely taught in universities. Saying that this is a response to environmental changes, necessarily means that first, there is a change in the environment, and then there is a response. And the response is defined as upright-walking.

> *There are many ideas about the role of the environment in human evolution. Some views assume that certain adaptations, such as upright walking or tool-making, were associated with drier habitat and the spread of grasslands, an idea often known as the*

> *savanna hypothesis. According to this long-held view, many important human adaptations arose in the African savanna or were influenced by the environmental pressure of an expanding dry grassland.* - Smithsonian Institution 45

The response to the change in the environment in the savanna theory was upright-walking.

Schombert and many other evolutionists assume that as the environment changes, DNA sequences also change to make organisms more adapted to their environment. But they don't explicitly say this. I'm unfolding this theory and explicitly stating the necessary step of changes to DNA. As DNA is what determines all the parts and functionality needed for an organism to walk uprightly as its primary mode of movement, this functionality would need to be encoded into DNA for it to show up in the phenotype.

Doubt has been cast on the savanna hypothesis, since it was first proposed, as we see in the following quote.

> *Numerous other explanations for bipedalism have been outright rejected, such as the idea that our ancestors needed to stand up to see over tall grass or to minimize the amount of the body exposed to the sun in a treeless savanna. Both ideas were debunked by the*

> *fact that the first hominids lived in at least partially wooded habitats.* Wayman, E. 53

However, there are new hypotheses and discoveries which haven't yet made it to the lectures in universities and print in textbooks. These also propose that changes in the environment are responsible for the rise of bipedalism in the predecessors to humans. We will look at a couple of these other explanations.

The first of these seems to be an adaptation of the savanna hypothesis.

> *Scientists think that these quick changes in climate caused our ancestors to become more flexible and resourceful. Bipedalism could have allowed our ancestors to be more flexible. Instead of only relying on the trees in wooded habitats, bipedal hominins could live in the forest and the grassland.* - Neysa Grider-Potter 21

This newer hypothesis suggests that the addition of grasslands to the forest, to make a mixed environment, caused bipedalism.

Here is another hypothesis that also ties bipedalism to changes in the environment, as we can see from this citation from a scientific paper by human evolutionary biologist Carsten Niemitz.

> *The forests were not far from a shore, where*

> *our early ancestor, along with its arboreal habits, walked and waded in shallow water finding rich food with little investment. In contrast to all other theories, wading behaviour not only triggers an upright posture, but also forces the individual to maintain this position and to walk bipedally. So far, this is the only scenario suitable to overcome the considerable anatomical and functional threshold from quadrupedalism to bipedalism.* 37

This hypothesis resonates with Lamarck's hypothesis of inheritance of acquired characters. It mentions "the considerable anatomical and functional threshold from quadrupedalism to bipedalism." But once again, the paper does not offer specifics of how information in the organism changes to overcome that threshold. The paper only mentions DNA in the title of one of its reference works. It only mentions mutations once.

Recall that at the beginning of this chapter, I said, "Lamarck and others of his time were making assumptions regarding heredity of which they didn't have even a basic knowledge?" *In regards to "the considerable anatomical and functional" changes that would be needed to enable our ancestors to become bipedal, we have the same lack of basic knowledge to back the aforementioned hypotheses just as Lamarck had a lack of knowledge to validate his assumptions.*

The Answer Has To Be In DNA To Be Valid

We understand how proteins are encoded in DNA. We know that proteins are the basic building blocks of all our anatomical and functional characteristics. What we don't know is where and how these anatomical and functional characteristics are encoded in DNA. We cannot go through DNA and point out the specific place within DNA where the attachment of a particular muscle to a bone is defined, where the width of a specific blood vessel is determined, or the definition of curvature and dimensions of each bone. Currently, scientists lack basic knowledge in regards to how anatomical and functional design are encoded in DNA.

"Nevertheless, our understanding of how the information encoded in a genome can produce a complex multicellular organism remains far from complete. To interpret the genome accurately requires a complete list of functionally important elements and a description of their dynamic activities over time and across different cell types. As well as genes for proteins and non-coding RNAs, functionally important elements include regulatory sequences that direct essential functions such as gene expression, DNA replication and chromosome inheritance." - Celniker, S, et al. 11

From the perspective of information technology, that information has to be somewhere, and we assume it is in DNA, and it has to be encoded in some form. We're far from understanding how this information would be encoded in the As, Ts, Cs, and Gs of DNA.

In spite of this lack of basic knowledge, hypotheses are proposed that involve environmental changes translating over and causing changes in DNA or *genotype,* which then cause changes in the organism or the phenotype.

Does the environment change DNA? We looked at this in the previous chapter. Yes, the environment changes DNA in random ways, but not in the way these hypotheses suggest. The environment does not make an organism able to walk upright and be better equipped to survive in the new environment. In chapter 8, we saw that environmental mutagens occur for physical reasons totally unrelated to how organisms survive in the environment. Environmental mutagens do not bring specific information about the environment to DNA. Mutations bring about changes that are random and not specific to the functionality of an organism.

To further evaluate the hypotheses proposing that changes in the environment change the organism, let's take a what-if look at what would be required to close the 'feedback loop' of environment changes and DNA changes. We will do it through the lens of information.

For our what-if analysis, let's first state the problem by defining what would need to happen if we wanted to see a response to a change in the environment. A straightforward "change in environment" problem statement can be: there is less room in the shrinking rainforest, and there is growing room in grasslands. Or perhaps, humanoids are now having to look for food in shallow water.

The second step is to consider possible solutions. One possible solution would be improved efficiency in quadrupedal motion. Another solution would be to swim like a dog in the shallow water. A third option, of obvious interest, would be bipedalism. Each of these solutions could be tied to the same problem statement, so there might be more than one solution to the change in the environment.

The third step would be to decide on a solution. In following with the hypotheses, let's respond by being able to walk upright. But how would this be decided in a natural setting?

The fourth step is to define the phenotype elements. The phenotype refers to the visible, useful, notable, observable characteristics of an organism. So you'd have to design the characteristics or phenotype elements required to walk upright as its primary mode of movement.

What are some of those elements?

Bones: You would need design changes to bones, maybe even add some new bones, maybe get rid of some bones. Specifically, we would need design changes to bones, to the spine, to the femoral, to the hip socket, the toes, feet, ankles, and neck. These changes in bones would be needed to walk upright as the primary mode of travel.

Legs would need a strong connection with the hip to support the upper body's weight. We would need strong knees, with broad areas below the knee, because at least briefly, only one leg would be carrying all the weight of the body. A curve in the lower back would help absorb the shock of going up and down as you walk. The skull would need to connect to the spine at a different angle, almost vertical, near the center of gravity of the head, so that the head is held upright. The upper leg bone would need to be longer to deal with the force of walking and running. 46

Muscles: As bones changed, muscles would also have to change, in synch. Changes would be necessary to the muscles in the feet, the legs, the back, the arms, the neck, and other places.

> *"Many researchers studying human fossils have focused on whether the foot was rigid or mobile but it's actually not about that, it's about when the spring-like motions in the foot were present."* 19

Interconnections: The bones would need spaces and connecting points for these muscles. You can't have a muscle in a given location if there's a bone there. You can't have a bone in a specific place if there's a muscle there. And those muscles have to hook onto the bones in good, leverageable ways if I could call them that. The ligaments, tendons, and connective tissues would also have to change in synch, with changes to the design of bones and muscles.

Nerves: Nerves would have to be rewired for there to be adequate feedback to the brain required for upright-walking. You need to feel where the ground is. You need to feel if you're stepping on your toe, or if you're on your heel. In the same way, you need to sense that you're moving forward or that you're not moving forward; sense that you're going left, or you're going right and so forth. So nerves would have to be rewired constantly for adequate feedback to the brain as a humanoid made a transition from quadrupedal walking to upright walking.

Circulatory System: You would need changes to the circulatory system so that sufficient blood is available for the new and expanded use of muscles. Maybe the muscle, that an ape wouldn't use much while moving on four legs, will now be used a lot to walk upright. He's going to use different muscles. Those muscles are going to need extra blood to get the oxygen and energy to them to work.

Dynamic Function: You need to make sure all this will function in all positions and actions. For example, it has to work when you're standing, and it has to function when you're sitting. It has to work when you're lying down. It has to function when you're walking. All this has to operate when you're running, and it has to work when you're stooping down. And all these organ systems —skeletal, muscular, nervous, circulatory— most work together every step of the process of incremental change.

Let me pause right there. For the sake of discussion, let's say we've settled all these functionality issues. Now I will propose a thought experiment and ask if a highly trained team of specialists could take an ape and make it primarily walk upright. Let's say you have an orthopedic unit and neurologists and experts from all of the necessary specialties. Could they do a series of operations on this ape, so that he not only could walk, but would primarily walk upright? Suppose we could provide all resources, even a small collection of male and female apes. Could they use gene editing to make it work? If we enabled them in any way possible, could they do it? The challenge we pose here is similar to the level of definition and implementation of functional elements that would be needed to be changed within DNA to solve our problem statement.

What I want to point out is the added complication that information in DNA brings to this problem. Thinking people way back in Darwin's time speculated that the environment directly changed animals, and that's the way they changed. Somehow or other, many have held on to that idea, even though we have a modern understanding of DNA.

What has DNA changed about it? We've briefly gone over the most obvious functionality issues. But it's a lot more complicated than this summary. I don't have a medical degree, but doctors and scientists with medical degrees could refer us to specialists for each one of those areas, and they would know the specific details involved in each area. And our understanding of the complexity involved would dramatically increase.

These are scientific hypotheses. They suggest that the environment changed, and then the apes responded by changes via evolution to walk upright as their primary mode of movement. Knowing what we know about DNA, all that new functionality in response to changes in the environment would have to be expressed and encoded into Ts, As, Cs, and Gs of the DNA of germ cells.

What if all these changes were just fortuitous? Let's ignore for a moment the statistical probability of having all these changes happen fortuitously. If all these changes happened fortuitously, we do not have a logical or scientific explanation for the information in DNA. The cause would be a lucky chance.

Since we are looking at this from an information perspective, I have several questions regarding this supposed change of DNA. It takes three nucleotides to specify one amino acid, as we've covered in one of the earlier chapters. One question is, how many nucleotides need to be added or changed to specify all this new functionality? I would speculate at least thousands or even tens of thousands of nucleotides would have to change to get this new upright walking functionality.

We know that it's just a what-if exercise. The environmental changes to DNA do not cause changes in genotype, which would cause changes in the organism's phenotype so that it is more adapted to the changing environment. We know that's not true. There are no flows of information through here that would make the organism more fitted to that environment. We saw that mutations are random and unrelated to the environment. We know that mutations bring no new information with them. They only modify existing information.

We can finish this what-if scenario. What we understand is that we have random mutations. These mutations cause random DNA changes in the genotype. It's not clear how anyone proposes that all these random changes cause upright-walking in the phenotype. Remember, what's driving modifications to DNA is not the environment changing. It's just random mutations.

We estimate that we would need tens of thousands of changes in its DNA sequence to modify an ape so that it adopted upright-walking. In the savanna hypothesis, environmental change is described as, "The rainforest was turning into steppes and grasslands due to global cooling." Is there anything about this environmental change that would induce an abnormal barrage of mutations in the DNA of germ cells of primates or apes? You could argue that perhaps venturing into steppes or grasslands would expose them to more radiation than in the rainforest, so possibly a small increase in mutations for that reason. Certainly not tens of thousands of additional mutations changing the functionality of bones, muscles, nerves, arteries, veins and other tissues.

On the other hand, Professor Schombert states that this environmental change, "drove two-thirds of their species into extinction." A decreasing population is going to produce less total germ cell mutations than a larger population. So it seems to be illogical to expect more changes in organisms during this period.

With our current scientific knowledge, these hypotheses proposing that the environment causes bipedalism, seem to be mere speculation. Why do I say that?

It is widely accepted that DNA contains all the information required for the development of the adult organism. We do not understand how this information is encoded and deciphered. We do not know how any structures above the level of proteins are specified within DNA. From an information perspective, we know that if the information is in DNA, it has to be in the abstract form of ACGT nucleotides.

Anyone who proposes a hypothesis stating that

changes to the environment cause changes to the functionality of an organism should also propose the specifics of how the environment would change DNA, at every step of the transformation, for such a beneficial outcome. Otherwise, these hypotheses are nothing more than unscientific musings.

Chapter 11
Does Natural Selection Change DNA?

In the previous two chapters, we examined whether environmental mutations or changes to the environment, can change the DNA of an organism enough to change how it functions. Now we ask another set of questions. Does natural selection change the sequence of nucleotides? Does natural selection add, take away, or move any nucleotides?

The simple answer is that natural selection does not cause changes in DNA sequences. We will discuss why that is the case in this chapter. First, let's define what we mean by natural selection.

> *For natural selection to operate, alleles that can occupy the same locus in the genome must differ somewhat between individuals. Such variation can appear because of replication or transcription errors, because of damage by radiation, or from other causes.*
> Steen, F. F. 48

Natural selection requires differences in the genome or DNA sequence. Natural selection, as its name implies, selects among the effects of these differences in organisms. We have discussed the causes of mutations in DNA. Natural selection does not cause mutations. Natural selection describes the process by which nature selects organisms based on their characteristics. An organism's characteristics are the combined result of the organism's unique variations of DNA.

Natural selection does not cause mutations. There is no feedback loop where natural selection affects how many mutations occur, what parts or functions the mutations affect, or whether they would be beneficial or not.

> *"Natural selection has no intentions or senses; it cannot sense what a species or an individual 'needs.' Natural selection acts on the genetic variation in a population, and this genetic variation is generated by random mutation — a process that is unaffected by what organisms in the population need."* 51

While natural selection does not guide or direct the quantity, type, or effects of mutations, it seeks to explain the non-random selection to which all living organisms are subject.

> *This is why "need," "try," and "want" are not very accurate words when it comes to explaining evolution. The population or individual does not "want" or "try" to evolve, and natural selection cannot try to supply what an organism "needs." Natural selection just selects among whatever variations exist in*

the population. The result is evolution. At the opposite end of the scale, natural selection is sometimes interpreted as a random process. This is also a misconception. The genetic variation that occurs in a population because of mutation is random — but selection acts on that variation in a very non-random way: genetic variants that aid survival and reproduction are much more likely to become common than variants that don't. Natural selection is NOT random! 52

Charles Darwin thought a lot about how humans select and breed plants and animals, seeking to get organisms that best fit their objectives. You could call breeding selection a result of human intervention or 'human selection.' Darwin proposed and pointed out that, without the input of humans, nature also selects organisms. He called this natural selection. So it is appropriate to draw some comparisons between human selection and natural selection.

We all are quite familiar with the great diversity of dogs around us. There are all different sizes and even variations in the relative size of their legs, necks, and heads. The shapes of their heads and the look of their faces vary significantly. Their energy levels and behavior are also different between one breed and another.

These differences are not accidental. They have been directed and sought by determined human breeders or selectors.

> *People have been breeding dogs since prehistoric times. The earliest dog breeders possibly selected wolves that had domesticated themselves to breed. From the beginning, humans purposefully bred dogs to perform various tasks. Hunting, guarding, and herding are thought to be among the earliest jobs eagerly performed by the animal destined to be called "man's best friend."* 3

If we look at the American Kennel Club's website, 3 we can see that dogs have been bred for purposes such as sporting, working, herding, sniffing, and even as toys. As a result of this breeding, some dogs are smarter, while others are equipped to guard people and properties. Some dogs are bred to be better with children or to live in apartments. There are breeds of hypoallergenic dogs or even hairless dogs.

How do breeders get dogs that have these specific traits? They start with dogs that have some of the characteristics they want and have them mate and produce offspring. Out of the progeny, they pick the puppies or dogs that show the desired characteristics most clearly. Then they breed these dogs, and they repeat the process. They do not mate the dogs that show less of the desired characteristics. They select among the offspring the dogs with the desired characteristics. If they would like to add some characteristics, not in the original group of dogs, then at some point, they may mate some of their select dogs with dogs that most clearly show the additional desired characteristics.

After many generations, breeders have rejected dogs that have genes that would produce dogs without the desired characteristics. If they do this enough, then they can speak of a purebred dog. What is meant by purebred? It means you can take a purebred male and a purebred female of the same type and mate them and be almost certain that their offspring will have the same characteristics as the parent dogs. This is because they have selected in favor of genes that produce these characteristics. So both male and female purebreds will have these genes. At the same time, they have selected against the genes that would interfere with the desired characteristics. The result is that purebreds do not have these undesired genes within them. The American Kennel Club has registered almost 200 different pure breeds.

Until recent advances that gave us the ability to read DNA, the breeders had no way of knowing what genes they were selecting for and what genes they were selecting against. Breeders don't need to know the genes. They simply look at the results.

From what we know of mutation rates, the mutations that have occurred since people started breeding dogs have probably contributed very little to the overall diversity of dog breeds. Up until recently, there was no way to edit DNA. Most of the breeding has just been selecting for or against genes that were already present in the wolves or other canines they included in their breeds. Human breeding did not introduce new genes. It just emphasized or exaggerated characteristics caused by existing genes.

All of this is in accord with the principles of genetics that Gregor Mendel worked out with his experiments with pea plants. Mendel bred pea plants until he had purebred pea plants, which could exhibit the specific characteristics he wanted.

This diversity is quite remarkable. Yet in accordance with Mendel's principals of genetics, if we were to take, five males and five females of each of the almost 200 pure breeds and let them freely breed amongst themselves, in a dozen or so generations we would begin to see a loss of their specialized traits. Soon you would have dogs that looked more like the original canines humans began to breed.

Natural selection, similar to human selection, can influence which dogs survive and multiply. This is especially true if there is not enough food or space in the environment for all the offspring to live and reproduce.

Continuing with our example, if we take the 2,000 or so dogs, which would be males and females of each of the purebreds and put them in an environment that presents challenges to living and multiplying, then we could see some natural selection in action. For example, if water were scarce, then the dogs that could survive on less water would be the most likely to survive and reproduce. Whereas if the environment were very frigid, then the dogs that would best live in the cold and their offspring would be more numerous than the dogs who die or are in poor health due to the cold.

> *Evolutionary fitness and success refers to surviving long enough to pass genetic material on to offspring. Traits that are passed on to offspring because they contribute to success are "selected for continuation." Traits that are eliminated from the population because they detract from success are "selected against continuation."* 13

So depending on the level of difficulty to survive and reproduce, we would get more dogs that are better fit for that specific environment. Here again, though, changes caused by mutations would be very few. Most of the selection would act upon the existing variations in the genes or DNA already present in dogs.

From our perspective of information, we see that since natural selection does not add or change nucleotide sequences, it does not add information to DNA.

Chapter 12
What Is Natural Selection Supposed To Do?

Our DNA question is, where does the information come from? We have briefly looked at the processes that cause mutations and have seen that none of them bring new information into DNA. Mutations do not answer the question of where the information comes from. Let's look then at natural selection. What is natural selection supposed to do? What do proponents of natural selection, as the explanation for the origin of all species, claim that it does?

> *The theory of natural selection has a big job —the biggest in biology. Its task is to explain how every adaptation evolved, step by step, from traits that preceded it. This includes not just body form and color, but the molecular features that underlie everything.* - Jerry Coyne 16

DNA mechanisms ensure that DNA is copied without errors and repaired if possible. Nevertheless, there will be random, unrelated, irrelevant, accidental, incidental, and possibly fortuitous mutations appearing as organisms reproduce. For it to work as evolutionary biologists claim it does, natural selection would and should filter out of this chaotic flow of mutations information, which is the source of functionality in organisms.

Chemists and biologists are quite familiar with filtering. Indeed filtering played a crucial role in arriving at our current understanding of DNA. Filtering DNA and learning the ratios of adenine (A) to thymine (T), as well as the ratio of cytosine (C) to guanine (G), helped us better understand the structure of DNA. Filtering is also used to produce the well-known DNA markers.

Information technology engineers are also familiar with filtering. Sometimes IT engineers clean up data, taking out what they consider noise. Other times the IT engineers do searches among random chunks of data to find the information within it.

Chemists, Biologists, and IT engineers build or adapt their filters based on what it is that they want to exclude, select, or retain. Natural selection only has one criterion that it uses to filter. The only test of natural selection for filtering is this: does this mutation raise this organism's effective reproducibility? If this mutation increases the effective reproducibility of the organism that has the mutation, then it reproduces more of its offspring. If this mutation lowers the effective reproducibility of the organism that has the mutation, then it reproduces less or has no offspring.

I refer to the organism, singular, that has the mutation because mutations occur initially in one single organism. While it's emphasized that populations evolve, mutations start in a single organism. They must increase that organism's and its offspring's effective reproducibility enough that its offspring eventually represent a significant part of the population.

Complementary Information

We have seen that information produces the functionality of organisms. If we were able to pick or design a filter to use for selection, the ideal filter would select for information. At least two complementary pieces of information are required for them to be useful in an information system, such as an organism with DNA. Often many more pieces of information are needed. A single piece of information by itself is not useful.

Before we look at an example in DNA of the need for two complementary pieces of information, let me give a simple hypothetical from everyday life. Let's take the case of a newly built house. The municipal government or some numbering organization will be in charge of assigning this new house a number. The house is built on Example Street, and the numbering organization informs the builder of the house that it will be assigned number 2374. The builder posts this number on the outside of the house on Example Street, and the address becomes 2374 Example Street. This piece of information, this house number is useless and non-functional until the new address is given to a person or organization who is looking for the house or wants to send mail to the house. So far, we are missing a second piece of information – we have a house with an address but no senders.

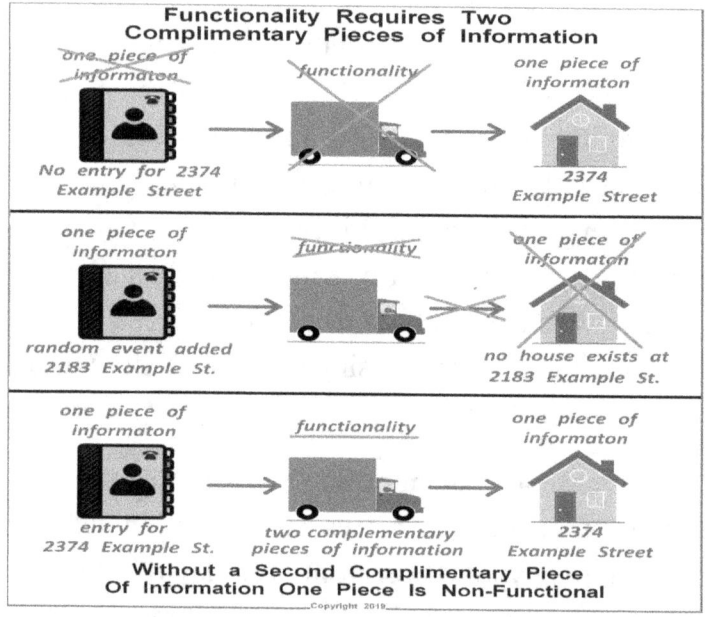

Complementary Pieces of Information [17]

What if we had a sender but were missing the other second complementary piece of information – the intended house. Say a random event changed an entry in an address book on Example Street to house number 2183, and there is no house with that number. If this address is given to the senders, the mail or package delivery is now sent to 2183 Example Street where there is no house. This piece of information is useless and non-functional, as in the first scenario. We have a sender, but because there is no house at the specified address, no deliveries will be completed. You need at least two valid and complementary pieces of information for each of them to be functional or useful.

Natural selection would not favor either of these random changes by themselves since individually they represent no improvement in functionality. For an organism to have an increase in reproducibility, there needs to be two or more pieces of complementary information that specify functionality. *However, natural selection would not favor an isolated piece of information without a second complementary piece of information because neither of the two pieces of information without the other improves functionality.* It would require two separate random events to change two separate pieces of information, such as the house number and the address book in our example. Natural selection cannot favor one of these random events because, hypothetically, in some future generation, there will be a second, random event that produces complementary change. *Thus natural selection cannot be a filter that selects two or more pieces of complementary information from the chaotic stream of DNA mutations.*

Complementary Information In DNA

Now let us look at an example of two required pieces of information in a functioning DNA system.

In the chapter Information in DNA, we saw that to build a protein, an enzyme called RNA polymerase is sent into the nucleus of the cell to transcribe the DNA sequence of the desired protein into what is called messenger RNA or mRNA. We began with a simple understanding of coding DNA. We will now go more in depth into how this works.

How does the RNA polymerase "know" which protein it is supposed to transcribe? In humans, there are more than 20,000 different sequences of DNA that each code for a different protein. How is the RNA polymerase supposed to know where in the strand of DNA it should start transcribing to get the protein that is needed?

For an RNA polymerase to be able to transcribe the contents of a protein in DNA into RNA, it has to know where the code for the protein begins on the DNA chain. There is a sequence of nucleotides upstream from the start of the protein called a *promoter*. Each promoter sequence is unique. There are proteins called transcription factors that identify the specific sequence of nucleotides in the promoter for the desired protein. The promoter's function is very much like the street number in our house analogy above. The promoter identifies which protein is encoded in the following DNA sequence.

> *In this review, we define the core promoter as the minimal stretch of contiguous DNA sequence that is sufficient to direct accurate initiation of transcription by the RNA polymerase II machinery. ... Typically, the core promoter encompasses the site of transcription initiation and extends either upstream or downstream for an additional ~35 nt.* [9]

As we see in the citation above, the promoter is a unique sequence of nucleotides (abbreviated nt) that identify where the coding information is for a specific protein. In many cases, the promoter can be readily located, as it will be close to a TATA box.

> *Transcription for protein-coding regions is often indicated by a TATA box, a sequence of nucleotides TATAA just prior to the actual encoding region. Locating a TATA box is what RNA "does" in finding proteins...* 39

115

In case you had any doubts about DNA containing information, the citation here is taken from a programming assignment in a bioinformatics course at Duke University.

Here is an image that shows the relationship between a promoter, TATA box, and a DNA sequence that codes for a protein.

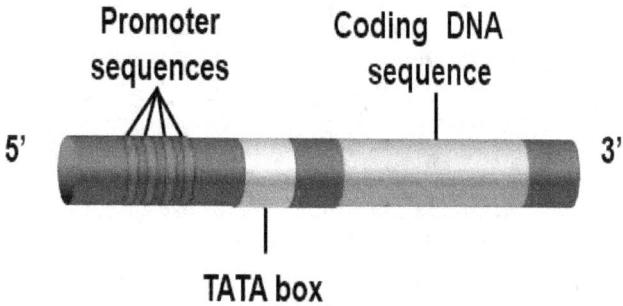

Promoter and Tata Box [16]

The promoter and the transcription factor are complementary pieces of information that have to match up to successfully transcribe the desired coding DNA sequence. One without the other is useless. One without the other does not allow transcription of any DNA sequence that follows, which may contain coding for a unique protein. In terms of natural selection, one without the other does nothing to increase the effective reproducibility of the organism. A mutation that changes the promoter but not the transcription factor is useless and vice-versa.

> ***Transcription has several important players that must all be in the right place at the right time: the transcription machinery, transcription factors, promoters, and enhancers. According to the existing model, transcription factors are proteins that bind to enhancer regions of the genome and recruit the transcription machinery to the promoter DNA regions, which then initiate the genes' transcription.*** *- Rura, N. G* 43

Again we see that two separate mutations corresponding to each other in two different locations would have to be in the DNA chain of the germ cells of an individual organism to get complementary information. It could be that one mutation was in the germ cell from the mother and the other from the father. If not, both mutations would have to be from the germ cell from either the father or the mother.

Why Natural Selection Fails The Complementary Information Test

The transcription factor is also a protein. We know from the central dogma that the transcription factor protein is likewise generated from coding sequences in DNA. To change the transcription factor so that it seeks a different promoter, there would have to be one or more mutations in the DNA coding sequence of the transcription factor protein. This would be the first required mutation.

At the same time, a different mutation would have to modify the sequence of nucleotides in the promoter upstream to match what this new transcription factor would seek. This would be the second required mutation.

Both of these changes would have to happen for a modified promoter, and it's transcription factor to create new functionality that could improve *reproducibility*. Only if there is improved effective reproducibility in this organism, would natural selection favor the organism.

If such a case as this were to occur, the new functionality would be produced only by chance. Natural selection played no part in setting up the two pieces of information. When we think about the 20,000 unique sequences of DNA that each code for a different protein in humans, it is not plausible to think that these sequences came from two unrelated random mutations that happened to functionally correspond to each other. These two mutations would have to show up in the germ cells of an individual organism to create a new functioning set of promoter and transcription factor.

So we can see that natural selection may be effective in selecting improvements in existing functionality that results in improved reproducibility. Natural selection would fail to select or favor changes in DNA that would be only the first change of two or more changes needed to set up a new functional information set. This is the DNA problem. We still do not know where that information comes from. Natural selection does not provide a satisfactory answer.

For natural selection to be the explanation of the origin of information, you would need to demonstrate some principles by which each *partial* piece of information contributes at least a slight advantage to the organism. At present, there is no basis to show that these *partial* pieces of information contribute anything to the effective reproducibility of any organism. This means that natural selection has no reason to select *partial* information to build information as that in DNA.

PART 3

Some Other Explanation?

Chapter 13
Do Epigenetics Change DNA?

So far, we've looked at whether changes to the environment, random mutations, and natural selection add new functional information to DNA, and I have argued they do not. It seems that many scientists agree we lack satisfactory explanations for how the information in DNA got there as they are now reaching for explanations that go outside the bounds of DNA. I am referring to epigenetics, which has gained prominence lately as a way to fill the gaps that remain when all other existing hypotheses are taken into consideration.

We've been examining these diverse external interactions, which some think would change DNA. As a reminder, when we consider the validity of the change in DNA that these interactions can bring, we specifically refer to a change in the sequence of nucleotides –adding to, subtracting from, or moving the order of the nucleotides.

To answer whether *epigenetics* change DNA, let's start by clarifying what epigenetics is. Here is a definition from an authoritative source.

> ***What is epigenetics? DNA modifications that do not change the DNA sequence can affect***

gene activity. Chemical compounds that are added to single genes can regulate their activity; these modifications are known as epigenetic changes. The epigenome comprises all of the chemical compounds that have been added to the entirety of one's DNA (genome) as a way to regulate the activity (expression) of all the genes within the genome. The chemical compounds of the epigenome are not part of the DNA sequence, but are on or attached to DNA ("epi-" means above in Greek). Epigenetic modifications remain as cells divide and in some cases can be inherited through the generations. 50

From this explanation, we understand that epigenetics do not change the sequence of A, G, C, and T nucleotides, how many there are, or their sequence order. Epigenetics does not change DNA nucleotides. Nevertheless, this definition states that epigenetic modifications "can be inherited through the generations," and thus, these modifications in some cases could be passed down to offspring.

Although epigenetics does not change DNA, it is useful for us to have some understanding of how epigenetics affects access to the information in DNA. To begin to understand epigenetics, we need to think about the length of DNA and the space it takes up. Consider how many nucleotides are in DNA:

> *The human genome contains approximately 3 billion of these base pairs, which reside in the 23 pairs of chromosomes within the nucleus of all our cells. Each chromosome contains hundreds to thousands of genes, which carry the instructions for making proteins. Each of the estimated 30,000 genes in the human genome makes an average of three proteins.*
> 49

In total, there are approximately three billion nucleotides. Even though they're just a few atoms in size, they do take up some space when there are three billion of them. The human genome, if it were stretched out, would be 2 meters, a little more than six feet, in length. 30

So you have this long DNA, which intuitively does not take up much space, as it's microscopic in width. But this long sequence has to fit into the nucleus of every cell. Think about the size of a cell and the size of the nucleus within the cell. This DNA has to fit into that nucleus of the cell. How can that much DNA fit in a compact space such as the nucleus of a cell?

We can think of this long DNA strand as a string. How would you store a long piece of string so it can be retrieved when needed?

Certainly, one option is to roll it onto a spool of

some type. If we thought that DNA could be stored that way, we would not be too far off, as pieces of DNA are rolled up onto a kind of spool.

> *Try holding a piece of string at one end, and twisting the other. As you add twist, the string creates coils of coils; and eventually, coils of coils of coils. Your DNA is arranged as a coil of coils of coils of coils of coils! This allows the 3 billion base pairs in each cell to fit into a space just 6 microns across.* 6

Here is an excellent illustration of how this coil of coils, of coils, looks on a microscopic level.

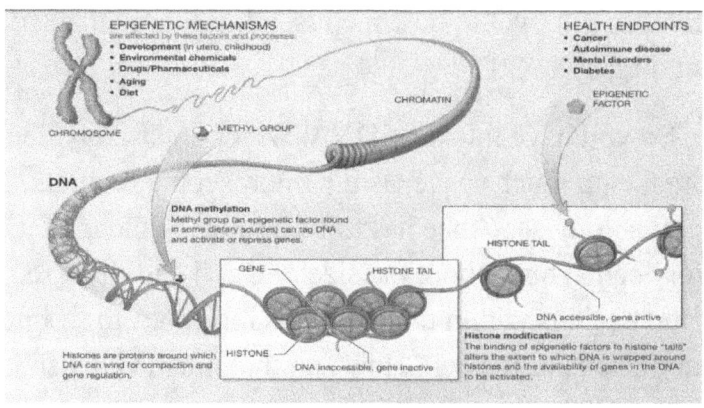

Epigenetic Mechanisms (14)

On the bottom of this illustration, you can see where the DNA strand is wrapped around *histone* proteins as if they were spools. Wrapping DNA around a histone protein may help organize the DNA strand. The histone also keeps the gene that is wrapped around it from being accessible, which makes it inactive. It's as if that DNA gene or sequence was not there because it's not being used in that particular epigenetic configuration. From an information perspective, histone proteins can make part of the information in DNA inaccessible unless other epigenetic factors intervene.

Here is another illustration of how DNA is wrapped around a histone protein, which is especially helpful from an information perspective. Suppose we had a book with lots of information, but then we took a big paperclip and clipped four or five pages together. As long as we didn't remove that paperclip, we could not read those pages of the book. That's a good illustration of epigenetics.

These histone proteins have tails. Other epigenetic factors can interact with these tails, which could cause the gene's DNA strand to unwrap from the histone protein and thus be accessible and active. Once the gene's DNA is unwrapped from the histone protein, it is active because then RNA polymerase can unzip that particular section of DNA and transcribe its DNA into RNA.

Below is a representation of DNA wrapped onto histone proteins.

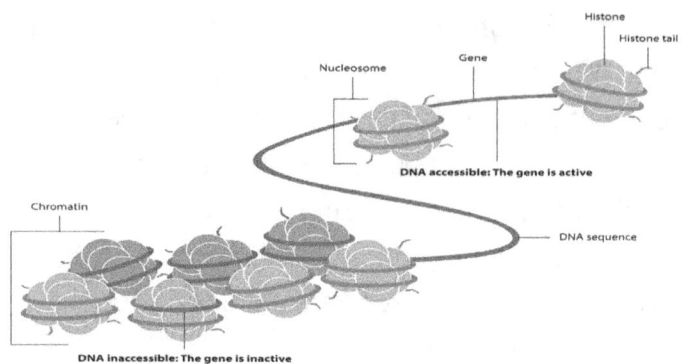

Epigenetic Histone Proteins (15)

Below is an explanation of epigenetics which expands on the definition I provided previously:

> *In recent decades, scientists have discovered that many traits in living things are controlled not just by their genetics, but also by processes outside their DNA that determine whether, when and how much the genes are expressed, known as epigenetics. This opens up the possibility of entirely new ways to breed plants and animals. By selectively turning gene expression on and off, breeders could create new varieties without altering the genes.* 32

Studies show that the environment in which the parent lived can cause epigenetic changes.

The flagship case study of the effects of epigenetics is based on what is known as the Dutch Hunger Winter. During WWII, Germany blockaded

Western Netherlands from the winter of 1944 through the spring of 1945. During this famine, the Dutch had to survive on less than a third of their required nutrition. Thousands died of malnutrition. Germany surrendered early in May of 1945, and soon after, Dutch nutrition levels recovered to close the recommended amount. The Dutch kept good medical records, and so researchers were later able to identify women who were pregnant during this time and follow up on their children and grandchildren. The effects varied depending on which part of their pregnancy they were malnourished. [10]

If they were malnourished early in their pregnancy but had proper nourishment in the final part of their pregnancy, their babies were born with regular birth weights. Later in life, they presented a tendency to be obese. This tendency was also passed on to their children.

If the mother had proper nourishment at the beginning of her pregnancy but was malnourished in the latter half, she generally would have small children. These children generally remained small for the rest of their lives. This tendency was also seen in their offspring.

Later, additional studies have confirmed that their children and grandchildren showed changes in their histone proteins. So it is understood that epigenetic changes caused these effects.

Back to our question, do epigenetics change DNA code? Do they add new nucleotides? Do they take away nucleotides? Do they change the order of nucleotides?

The answer is no, no, and no. Epigenetics does not change DNA, but rather, the change is in DNA expression or activation. Many detailed statistical studies have shown that changes in DNA expression are seen in the organism. But they have not demonstrated that new information is added to the DNA sequence. That answers our question. Epigenetics do not change DNA.

Let's note a couple more things. Histone proteins, like all other proteins, are encoded in DNA nucleotides. So the functionality of histone proteins comes from the information in DNA.

Epigenetics adds another level of complexity to using the information in DNA. It may be necessary for epigenetic agents to interact with the histone proteins before a given gene can be transcribed. Epigenetic agents are also proteins, which are encoded in nucleotide sequences in DNA. Information in DNA gives epigenetic agents their functionality.

We can see that not only does DNA contain information, but also that this information is built into a complex interdependent information system in the organism. Each piece of information is functional because of one, or most often more, complementary pieces of information. This system is needed to extract this information and use it to build and maintain the organism and its various functionalities.

Epigenetics doesn't add information to DNA. It can turn information in DNA on or off, but it doesn't create new information. Therefore, epigenetics is also an unsatisfactory explanation of where the information comes from.

Conclusion

We have to marvel when thinking of one cell multiplying in just nine months to a newborn baby who has all the structure and functionality of every organ, the organ systems, and the interplay between these. And the cells in that newborn baby contain all the information they need to continue to multiply in the future so that the baby grows and develops into an adult human being. And we know as humans that we are much more than just our physical bodies. So we cannot help but ask the question, where does it all come from?

We have seen that none of the proposed sources for the information in DNA send any information into DNA. We looked at mutations, the environment, and natural selection as possible sources and ruled these out as sources of the information. We then considered epigenetics as a source of information and found that it too, cannot add information. For this reason, we have to say that we do not know where the information in DNA originates.

We also do not know where within DNA is the information that would specify elements larger than proteins. We do not understand how organisms as diverse as a bird, a turtle, or an elephant and as unique as a human are built from similar pools of proteins.

In the preface, I shared a quote by Nobel Prize molecular biologist Jaques Monod on how the theory of evolution has shaped modern philosophical, religious, and political thought. Monod was an outstanding scientist. I would point out, though, that in his book "Chance And Necessity," he steps out of the strict bounds of material science to expound on his philosophical reflections on chance and evolution.

It has been my intent in this book to stay within the strict bounds of material science. However, in this conclusion, I, like Monod, will briefly jump out of those bounds to make a couple of observations.

First, Monod felt the need to write about chance as he realized that the theory of evolution depended on an enormous amount of very lucky randomness or chance. In his book, he almost seems to praise the powerful role of chance. It's not surprising that he would do this since the theory of evolution by natural selection is not a sufficient explanation for how life originated. By necessity, he attempts to explain the extremely lucky chance natural selection proposes.

If the origin of life is by chance and needs a philosophical explanation because science itself is insufficient, then there are undoubtedly other philosophical and theological perspectives to consider. Once we've established that science itself does not offer a satisfactory answer to the DNA Question, we should consider other explanations. Like Monod, we might do well to go outside of the strict bounds of material science in search of a logical answer.

Evolution by means of natural selection may be the best theory scientists have up until now. It is false to propose that it explains the origin of information in DNA and thus the origin of man and all organisms. Natural selection is supposed to explain function, but natural selection only operates when there is already functionality. There is no explanation for the origin of the information needed for each new functionality.

Good science requires us to reject explanations with little evidence and continue to search for a source for the information in DNA. Good science demands we continue to seek to understand how and where structural information is encoded in DNA.

We should conduct our science and our lives based on not knowing the origin of information in DNA.

I am amazed at the diversity, functionality, beauty, and joy of life. The more I have studied DNA, genetics, and information, my sense of awe has increased. The DNA Question has gotten more interesting, not less.

If you would like more information on this project.

www.DNAInformationResearch.com

www.DNAInformationResearch.com

Coming in 2020 the next book in this series.

the DNA problem

Natural Selection Does Not Account For
The Information In DNA
A deeper dive into information systems and DNA

Footnotes

1. Alberts, B., Johnson, A., Lewis, J., Raff, M., Roberts, K., & Walter, P. (2019). From DNA to RNA. Retrieved from https://www.ncbi.nlm.nih.gov/books/NBK26887/ ««
2. ArborGen Inc. (2019, July 13). Best Loblolly Pine Tree Seedlings: Forestry Supplier ArborGen. Retrieved July 13, 2019, from https://www.arborgen.com/what-are-your-seedling-choices/ «« 3. American Kennel Club. (n.d.). Dog Breeds. Retrieved July 27, 2019, from https://www.akc.org/dog-breeds/ ««
4. Bacon R. Opus maius. MS Digby 325, 15th century manuscript. Bodleian Library, Oxford, 1266 [translations consulted: Bridges JH, Oxford: Oxford University Press, 1897 and Burke RB, Philadelphia: University of Pennsylvania Press, 1928] ««
5. Bar-Yam, S. (n.d.). What Lamarck Believed - New England Complex Systems Institute. Retrieved September 8, 2019, from https://necsi.edu/what-lamarck-believed ««
6. BBC Science Focus. (n.d.). How long is your DNA? Retrieved July 4, 2019, from https://www.sciencefocus.com/the-human-body/how-long-is-your-dna/ ««
7. Bio.libretexts.org. (2018). 3.4: Proteins - Biology

LibreTexts. [online] Available at: https://bio.libretexts.org/TextMaps/Introductory_and_General_Biology/Book:_General_Biology_(OpenStax)/1:_The_Chemistry_of_Life/3:_Biological_Macromolecules/3.4:_Proteins [Accessed 22 Aug. 2018]. ««

8. Burkhardt, R. W. (2013). Lamarck, Evolution, and the Inheritance of Acquired Characters. Genetics, 194(4), 793–805. https://doi.org/10.1534/genetics.113.151852 ««

9. Butler, J. E. F., & Kadonaga, J. T. (2002, October 15). The RNA polymerase II core promoter: a key component in the regulation of gene expression. Retrieved August 12, 2019, from http://genesdev.cshlp.org/content/16/20/2583.long ««

10. Carey, N. (2012). Beyond DNA: Epigenetics | Natural History Magazine. Retrieved July 6, 2019, from http://www.naturalhistorymag.com/features/142195/beyond-dna-epigenetics ««

11. Celniker, S, et al (2009) Unlocking the secrets of the genome, Nature Publishing Group, http://www.nature.com/nature/journal/v459/n7249/full/459927a.html (accessed: June 06, 2014). ««

12. Cold Spring Harbor Laboratory. (n.d.). How insulin is made using bacteria: DNA Learning Center.

Retrieved July 9, 2019, from https://www.dnalc.org/view/15928-How-insulin-is-made-using-bacteria.html ««

13. College of Education, University of Hawai'i. (n.d.). Evolution by Natural Selection - Exploring Our Fluid Earth. Retrieved July 31, 2019, from https://manoa.hawaii.edu/exploringourfluidearth/biological/what-alive/evolution-natural-selection ««

14. Computational Medicine Center, DNA and RNA. (2019). Retrieved from https://cm.jefferson.edu/learn/dna-and-rna/ ««

15. Cotton Australia. (n.d.). Uses of Cotton | Cotton Australia. Retrieved July 12, 2019, from https://cottonaustralia.com.au/australian-cotton/basics/uses-of-cotton ««

16. Coyne, J. A. (2010). Why Evolution is True. Retrieved from https://books.google.com/books?id=dPh0DgAAQBAJ ««

17. Dawkins, R. (n.d.). The Information Challenge. [online] Australian Skeptics Inc. Available at: https://www.skeptics.com.au/resources/articles/the-information-challenge/ [Accessed 22 Aug. 2018]. ««

18. DiVenere, V., & Columbia University. (2017). Evolution. Retrieved September 2, 2019, from http://www.columbia.edu/%7Evjd1/evolution.htm ««

19. Gearhart Levy, R. (2017, February 18). The Evolution of Walking | Box | NYIT. Retrieved

September 6, 2019, from https://www.nyit.edu/box/features/the_evolution_of_walking «»
20. Genentech, Inc. (1978, September 6). Genentech: Press Releases | Wednesday, Sep 6, 1978. Retrieved July 9, 2019, from https://www.gene.com/media/press-releases/4160/1978-09-06/first-successful-laboratory-production-o «»
21. Grider-Potter, N. (2016, February 3). Ask An Anthropologist. Retrieved September 6, 2019, from https://askananthropologist.asu.edu/stories/walking-upright-tale-two-legs «»
22. International Service for the Acquisition of Agri-biotech Applications (ISAAA). (n.d.). Biotech/GM Trees | ISAAA.org. Retrieved July 13, 2019, from https://www.isaaa.org/resources/publications/pocketk/50/default.asp «»
23. LaMorte, W. W., "DNA, Genetics, And Evolution". 2018. Sphweb.Bumc.Bu.Edu. Accessed August 27 2018. http://sphweb.bumc.bu.edu/otlt/MPH-Modules/PH/DNA-Genetics/DNA-Genetics_print.html. «»
24. Lappin TRJ, Grier DG, Thompson A, Halliday HL. HOX genes: seductive science, mysterious mechanisms. The Ulster Medical Journal. 2006;75(1):23–31. http://www.ncbi.nlm.nih.gov/pmc/articles/PMC1891803 (accessed: June 08, 2014). «»

25. Levin, M. May 23, 2012. Dr. Michael Levin Investigator Seminar. Emergent Behaviors of Integrated Cellular Systems. http://ebics.net/events/dr-michael-levin-investigator-seminar (accessed: November 15, 2014) ««

26. Merriam-Webster.Com, Definition Of ABSTRACT. 2018. Accessed September 3 2018. https://www.merriam-webster.com/dictionary/abstract. ««

27. Merriam-Webster.Com, Definition Of BIOLOGY. 2018. Accessed August 26 2018. https://www.merriam-webster.com/dictionary/biology. ««

28. Merriam-Webster. (n.d.-a). Definition of DATA. Retrieved November 3, 2019, from https://www.merriam-webster.com/dictionary/data ««

29. Merriam-Webster. (n.d.). Definition of INFORMATION. Retrieved November 3, 2019, from https://www.merriam-webster.com/dictionary/information ««

30. Milo, R., Jorgensen, P., Moran, U., Weber, G., & Springer, M. (2010, January 1). BioNumbers—the database of key numbers in molecular and cell biology. Retrieved July 5, 2019, from https://www.ncbi.nlm.nih.gov/pmc/articles/PMC2808940/ Details at: https://bionumbers.hms.harvard.edu/bionumber.aspx?

id=108957 «««

31. Monod, J. (1972). Chance and necessity: an essay on the natural philosophy of modern biology (First Vintage Books Edition). New York, NY USA: Vintage Books. «««

32. National Institute of Environmental Health Sciences (NIEHS). (n.d.). Multimedia Gallery - First step toward epigenetically modified cotton (Image 2) | NSF - National Science Foundation. Retrieved June 29, 2019, from https://www.nsf.gov/news/mmg/mmg_disp.jsp?med_id=133150 «««

33. National Library of Medicine (NLM). (n.d.). What is noncoding DNA? Retrieved June 10, 2019, from https://ghr.nlm.nih.gov/primer/basics/noncodingdna «««

34. National Library of Medicine - Genetics Home Reference. (2019, October 1). What are single nucleotide polymorphisms (SNPs)? Retrieved October 10, 2019, from https://ghr.nlm.nih.gov/primer/genomicresearch/snp «««

35. National Research Council. 2005. Mathematics and 21st Century Biology, 3 - Understanding Molecules. Washington, DC: The National Academies Press. https://doi.org/10.17226/11315. «««

36. Niederhuber, M. (2015, August 11). Insecticidal

Plants: The Tech and Safety of GM Bt Crops - Science in the News. Retrieved July 12, 2019, from http://sitn.hms.harvard.edu/flash/2015/insecticidal-plants/ ««

37. Niemitz, C. (2010). The evolution of the upright posture and gait—a review and a new synthesis. Naturwissenschaften, 97(3), 241–263. https://doi.org/10.1007/s00114-009-0637-3 ««

38. Nuffield Council on Bioethics. (n.d.). The use of genetically modified crops in developing countries. Retrieved from http://nuffieldbioethics.org/wp-content/uploads/GM-Crops-Discussion-Paper-2003.pdf ««

39. Owen L. Astrachan, O. L., & Duke University. (2007, November 8). APT, Promotion and tata boxes I. Retrieved August 16, 2019, from https://www2.cs.duke.edu/csed/algoprobs/dna5-5.html ««

40. Pearson, G., & Oregon State University. (2019). BB331 Introduction to Molecular Biology, Lecture 8. Retrieved September 27, 2019, from http://oregonstate.edu/instruction/bb331/lecture09/lecture09.html««

41. Public Affairs (2019). Visualizing the central dogma - Cold Spring Harbor Laboratory. [online] Cold Spring Harbor Laboratory. Available at: https://www.cshl.edu/visualizing-the-central-dogma/

[Accessed 22 May 2019]. «««
42. Rocha-Munive, M. G., Soberón, M., Castañeda, S., & Niaves, E. (1970, January 1). Evaluation of the Impact of Genetically Modified Cotton After 20 Years of Cultivation in Mexico. Retrieved July 12, 2019, from https://www.ncbi.nlm.nih.gov/pmc/articles/PMC6023983/ «««
43. Rura, N. G., & Whitehead Institute - MIT. (2017, December 7). Rethinking transcription factors and gene expression. Retrieved August 18, 2019, from http://news.mit.edu/2017/mit-rethinking-transcription-factors-in-gene-expression-1207 «««
44. Schombert, J. (n.d.). 21st Century Science, Lecture 9, Evolution. Retrieved June 28, 2019, from http://abyss.uoregon.edu/%7Ejs/21st_century_science/lectures/lec09.html «««
45. Smithsonian Institution. (2018, September 14). Climate Effects on Human Evolution. Retrieved September 6, 2019, from http://humanorigins.si.edu/research/climate-and-human-evolution/climate-effects-human-evolution «««
46. Smithsonian Institution. (2018b, October 17). Walking Upright. Retrieved September 6, 2019, from http://humanorigins.si.edu/human-characteristics/walking-upright «««
47. Spencer , C., & Georgia Tech. (2018, September

1). 1.05 What is evolution and why do biologists think it's important? | The Biology of Sex and Death (BIOL 1220). Retrieved July 26, 2019, from http://bio1220.biology.gatech.edu/?page_id=254 «»«

48. Steen, F. F., & University of California, Los Angeles. (2001, March 25). Evolutionary Theory. Retrieved July 31, 2019, from http://cogweb.ucla.edu/ep/Evolution.html «»«

49. The National Human Genome Research Institute (NHGRI). (n.d.). Human Genome Project FAQ. Retrieved July 4, 2019, from https://www.genome.gov/human-genome-project/Completion-FAQ «»«

50. The National Institutes of Health, Roadmap Epigenomics Project. (n.d.). What is epigenetics? Retrieved June 29, 2019, from https://ghr.nlm.nih.gov/primer/howgeneswork/epigenome «»«

51. University of California Berkeley, Misconceptions about evolution. (2019). Retrieved from https://evolution.berkeley.edu/evolibrary/misconceptions_faq.php#b2 «»«

52. University of California Berkeley - Understanding Evolution. (n.d.). Misconceptions about natural selection. Retrieved July 24, 2019, from https://evolution.berkeley.edu/evolibrary/article/evo_32 «»«

53. Wayman, E. (2012, August 6). Becoming Human:

The Evolution of Walking Upright. Retrieved September 6, 2019, from https://www.smithsonianmag.com/science-nature/becoming-human-the-evolution-of-walking-upright-13837658/ ««

54. Wikipedia contributors. (2019, June 8). Noisy data - Wikipedia. Retrieved November 3, 2019, from https://en.wikipedia.org/wiki/Noisy_data ««

55. Zhang, Y., & Gladyshev, V. N. (2007). High content of proteins containing 21st and 22nd amino acids, selenocysteine and pyrrolysine, in a symbiotic deltaproteobacterium of gutless worm Olavius algarvensis. Nucleic acids research, 35(15), 4952–4963. doi:10.1093/nar/gkm514 ««

Glossary

4-dimensional three dimensions of space plus time dimension (7) ««

adenine a purine base; one of the four molecules containing nitrogen present in the nucleic acids DNA and RNA; designated by letter A. (4) ««

allele An allele is one of two or more versions of a gene. (5) ««

amino acids a set of 20 different molecules used to build proteins. (3) ««

antibody An antibody is a protein component of the immune system that circulates in the blood, recognizes foreign substances like bacteria and viruses, and neutralizes them. (5) ««

bacteria small single-celled organisms (3) ««

base One of the molecules that form DNA and RNA molecules. (2) ««

base pair two chemical bases bonded to one another, forming a "rung of the DNA ladder." (3) ««

binary Numbering system that only has two digits, sequences of 0's and 1's known as bits (7) ««

bioinformatics a subdiscipline of biology and computer science concerned with the acquisition, storage, analysis, and dissemination of biological data, most often DNA and amino acid sequences (3) «««

biotech Biotechnology, a set of biological techniques developed through basic research and now applied to research and product development. (4) «««

bits 0's or 1's used in binary numbering system (7) «««

Cell the basic building block of living things. (3) «««

chromosome an organized package of DNA found in the nucleus of the cell. (3) «««

codon a trinucleotide sequence of DNA or RNA that corresponds to a specific amino acid. (3) «««

cytosine a nitrogenous base, one member of the base pair GC (guanine and cytosine) in DNA. (2) «««

DNA Molecule that encodes genetic information, deoxyribonucleic acid (2) «««

effective reproducibility the ability of an organism to reproduce offspring that can also survive to reproduce (7) «««

enzyme a biological catalyst and is almost always a protein. It speeds up the rate of a specific chemical reaction in the cell. (3) «««

epigenetics an emerging field of science that studies heritable changes caused by the activation and deactivation of genes without any change in the underlying DNA sequence of the organism. (3) ««

epigenome derived from the Greek word epi, which literally means "above" the genome. The epigenome consists of chemical compounds that modify, or mark, the genome in a way that tells it what to do, where to do it, and when to do it. (3) ««

gene used vaguely as a basic physical unit of inheritance. Often used precisely to refer to DNA that codes for a protein. (3) ««

genetics The study of the patterns of inheritance of specific traits. (4) ««

genome the entire set of genetic instructions found in a cell. (3) ««

genotype an individual's collection of genes. The term also can refer to the two alleles inherited for a particular gene. (3) ««

germline the sex cells (eggs and sperm) that are used by sexually reproducing organisms to pass on genes from generation to generation. (3) ««

guanine one of four chemical bases in DNA (3) ««

hemoglobin iron-containing oxygen-transport metalloprotein in the red blood cells of almost all vertebrates (1) ««

heredity the transmission of characteristics from one generation to the next (4) ««

histone A histone is a protein that provides structural support to a chromosome. In order for very long DNA molecules to fit into the cell nucleus, they wrap around complexes of histone proteins, giving the chromosome a more compact shape. (5) ««

locus A locus is the specific physical location of a gene or other DNA sequence on a chromosome, like a genetic street address (5) ««

microbiology is the study of microorganisms (1) ««

mutagen a mutagen is a chemical or physical phenomenon, such as ionizing radiation, that promotes errors in DNA replication. (5) ««

mutation any change in a specific DNA sequence. (6) ««

nucleobase (or just base) nitrogen-containing biological compounds that form nucleosides, which in turn are components of nucleotides, (1) ««

nucleotide the basic building block of nucleic

acids. RNA and DNA are polymers made of long chains of nucleotides. (6) ««

nucleus a membrane-bound organelle that contains the cell's chromosomes. (3) ««

organ a collection of tissues that structurally form a functional unit specialized to perform a particular function. (3) ««

Paneth cells are a principal cell type of the small intestine (1) ««

polymerase an enzyme that makes RNA using DNA as a template in a process called transcription. (6) ««

peptide one or more amino acids linked by chemical bonds. (3) ««

phenotype functional and structural characteristics of an organism (6) ««

polymorphism one of two or more variants of a particular DNA sequence. The most common type of polymorphism involves variation at a single base pair. (3) ««

polypeptide a protein or part of a protein made of a chain of amino acids joined by a peptide bond. (2) ««

promoter a sequence of DNA needed to turn a gene on or off. The process of transcription is initiated at the promoter. (3) ««

protein an important class of molecules found in all living cells. A protein is composed of one or more long chains of amino acids (7) ««

pyrrolysine is an a-amino acid that is used in the biosynthesis of proteins in some methanogenic archaea and bacteria; it is not present in humans. (1) ««

recombinant DNA (rDNA) is a technology that uses enzymes to cut and paste together DNA sequences of interest. (5) ««

reproducibility the ability of an organism to reproduce (7) ««

RNA Ribonucleic acid (RNA) is a molecule similar to DNA. Unlike DNA, RNA is single-stranded. An RNA strand has a backbone made of alternating sugar (ribose) and phosphate groups. (3) ««

selenocysteine is the 21st proteinogenic amino acid, is a building block of selenoproteins. (1) ««

somatic cell any cell of the body except sperm and egg cells. (1) ««

stem cell A stem cell is a cell with the potential to form many of the different cell types found in the body. (5) ««

thymine one of four chemical bases in DNA (3) ««

transcription is the process of making an RNA copy of a gene sequence. (5) ««

translocate a large segment of one chromosome breaks off and attaches to another chromosome. (2) ««

uracil is one of four chemical bases that are part of RNA. (5) ««

valine is an a-amino acid that is used in the biosynthesis of proteins. (1) ««

virus an infectious agent that occupies a place near the boundary between the living and the nonliving. It is a particle much smaller than a bacterial cell, consisting of a small genome of either DNA or RNA surrounded by a protein coat. (3) ««

zygote a eukaryotic cell formed by a fertilization event between two gametes. (1) ««

Sources

(1) wikipedia.org

(2) genomicscience.energy.gov/glossary

(3) rarediseases.info.nih.gov/glossary

(4) strbase.nist.gov/glossary.htm

(5) www.genome.gov/genetics-glossary

(6) www.ncbi.nlm.nih.gov/probe/docs/glossary/

(7) Author's definition

Image and Illustration Credits

Cover Illustration, "DNA as an information container" Mayra Cañas, "Smartec Systems" (smartecsystemsc@gmail.com) ««

(1) T Cell NIH Intramural Research Program. (2018). Intermural Blog | NIH Intramural Research Program [Photograph]. Retrieved from https://irp.nih.gov/sites/default/files/blog/T+cell.jpg ««

(2) DNA Like a Twisted Ladder Courtesy: National Human Genome Research Institute, www.genome.gov ««

(3) DNA Base Pairs Courtesy: National Human Genome Research Institute, www.genome.gov ««

(4) DNA Codon Table Wikipedia contributors, "DNA codon table," Wikipedia, The Free Encyclopedia, https://en.wikipedia.org/w/index.php?title=DNA_codon_table (accessed May 17, 2019). ««

(5) Kras Protein Shape National Institute of Health - Image Gallery. (2017). Images and B-roll [Illustration]. Retrieved from https://www.nih.gov/sites/default/files/styles/featured_media_breakpoint-large-extra/public/news-events/news-releases/2016/20161208-nci.jpgv ««

(6) **DNA to mRNA** MRNA-interaction.png. (2007). Retrieved from https://en.wikipedia.org/wiki/File:MRNA-interaction.png «««

(7) **Central Dogma** Horspool, D. (2008). File:Central Dogma of Molecular Biochemistry with Enzymes.jpg - Wikimedia Commons. [online] Commons.wikimedia.org. Available at: https://commons.wikimedia.org/wiki/File:Central_Dogma_of_Molecular_Biochemistry_with_Enzymes.jpg [Accessed 24 May 2019]. «««

(8) **Synthetic Insulin** National Library of Medicine (NLM). (n.d.-a). How did they make insulin from recombinant DNA? Retrieved July 9, 2019, from https://www.nlm.nih.gov/exhibition/fromdnatobeer/exhibition-interactive/recombinant-DNA/alternative/recombinant-dna-technology-03.jpg «««

(9) **Missense mutation** National Library of Medicine (NLM). (2019). Missense mutation [Illustration]. Retrieved from https://ghr.nlm.nih.gov/primer/illustrations/missense.jpg «««

(10) **Deletion mutation** National Library of Medicine (NLM). (2019). Deletion mutation [Illustration]. Retrieved from

https://ghr.nlm.nih.gov/primer/illustrations/deletion.jpg ««

(11) Insertion Mutation National Library of Medicine (NLM). (2019). Insertion mutation [Illustration]. Retrieved from https://ghr.nlm.nih.gov/primer/illustrations/insertion.jpg ««

(12) DNA mutation caused by UV light NASA's Earth Observatory. (2001). Ultraviolet (UV) radiation [Photograph]. Retrieved from https://earthobservatory.nasa.gov/ContentFeature/UVB/Images/dna_mutation.gif ««

(13) Types of Virus National Human Genome Research Institute (NHGRI). (n.d.). Types of Virus [Illustration]. Retrieved from https://www.genome.gov/genetics-glossary/Virus ««

(14) Epigenetic Mechanisms National Institutes of Health. (n.d.). Epigenomics - Epigenetic Mechanisms. Retrieved June 29, 2019, from https://commonfund.nih.gov/epigenomics/figure ««

(15) Epigenetic Histone Proteins National Institute of Environmental Health Sciences. (n.d.). Epigenetics may hold key to arsenic's role in cancer (Environmental Factor, January 2019). Retrieved June 29, 2019, from https://factor.niehs.nih.gov/2019/1/science-highlights/epigenetics/index.htm ««

(16) Promoter and Tata Box Luttysar, ?, & Wikimedia Commons. (2017, October 9). File:TATA box structural elements.png - Wikimedia Commons. Retrieved August 16, 2019, from https://commons.wikimedia.org/wiki/File:TATA_box_structural_elements.png ««

(17) Complementary Pieces of Information Searfoss S. - All rights reserved««

Additional Resources

We will post additional resources on our webpage The DNA Question. The webpage will provide links for resources to learn more about DNA and information. There will also be a FAQ section as we receive feedback from the book. Here is the link to our webpage.

www.theDNAquestion.com

If you would like to get involved in our work to investigate and teach regarding information and DNA, we invite you to visit our non-profit organization's website. Here is the link to the DNA Information Research site.

www.DNAInformationResearch.com

Appendix - Abstract Information

How to make Melcena-Deborah Searfoss brownies «««

Ingredients list:

   2 cups sugar
   1/2 cup cocoa powder
   1 cup corn oil
   4 eggs
   1 tsp vanilla
   1 tsp salt
   1 tsp baking powder
   1 1/4 cup flour

Instructions:

   1 - mix the dry sugar and cocoa powder until evenly mixed
   2 - add the flour to the sugar and cocoa mix
   3 - Using mixer, combine oil, eggs, vanilla, salt, baking powder
   4 - Add sugar, cocoa powder, and flour to liquid ingredients
   5 - Mix combined ingredients
   6 - Butter a 9" x 12" pan and add mix
   7 - Bake at 350 F. for 25 minutes

www.ingramcontent.com/pod-product-compliance
Lightning Source LLC
Chambersburg PA
CBHW060841220526
45466CB00003B/1191